Geology of the country around Lewes

The Lewes district is one of contrasting scenery, including parts of the High Weald, the South Downs and the intervening lowland. These tracts are dissected by the valleys of the River Ouse and the Cuckmere River, which form impressive gaps in the chalk downland. In the south-east the reclaimed marshland of the Pevensey Levels extends over a wide area near the coast.

The strata at the surface are mainly of Cretaceous age and range from the lower part of the Purbeck Beds to the Upper Chalk. This stratigraphical account describes the various formations present and, in particular, provides details of a refined stratigraphy for some of the Wealden formations, the Gault and the Chalk. These details are largely based on cored borehole sequences and on carefully measured sections in chalk-pits.

Plate 1   Chalk scarp of the South Downs near Alfriston (A 12306)

BRITISH GEOLOGICAL SURVEY

Natural Environment Research Council

R. D. LAKE,
B. YOUNG,
C. J. WOOD and
R. N. MORTIMORE

Geology of the country around Lewes

Memoir for 1:50 000 geological sheet 319
(England & Wales)

CONTRIBUTORS

Stratigraphy
R. A. B. Bazley
C. R. Bristow

Palaeontology
F. W. Anderson
D. E. Butler
B. M. Cox
H. C. Ivimey-Cook
A. A. Morter

Petrology
R. J. Merriman

Water Supply
R. A. Monkhouse

LONDON: HER MAJESTY'S STATIONERY OFFICE   1987

© *Crown copyright 1987*

First published 1987

ISBN 0 11 884406 7

Bibliographical reference

LAKE, R. D., YOUNG, B., WOOD, C. J. and MORTIMORE, R. N. 1987. Geology of the country around Lewes. *Mem. Br. Geol. Surv.*, Sheet 319 (England and Wales), 117 pp.

Authors

R. D. LAKE, MA and C. J. WOOD, BSc
British Geological Survey, Keyworth, Nottingham NG12 5GG

B. YOUNG, BSc
British Geological Survey, Windsor Court, Windsor Terrace, Newcastle upon Tyne NE2 4HB

R. N. MORTIMORE, BSc, PhD
Brighton Polytechnic, Moulsecoomb, Brighton BN2 4GJ

Contributors

B. M. Cox, BSc, PhD, H. C. Ivimey-Cook, BSc, PhD and R. J. Merriman, BSc
British Geological Survey, Keyworth, Nottingham NG12 5GG

R. A. B. Bazley, BSc, PhD
British Geological Survey, Bryn Eithyn Hall, Llanfarian, Aberystwyth, Dyfed SY23 4BY

C. R. Bristow, BSc, PhD
British Geological Survey, St Just, 30 Pennsylvania Road, Exeter EX4 6BX

R. A. Monkhouse, MSc
British Geological Survey, Maclean Building, Crowmarsh Gifford, Wallingford OX10 8BB

The late F. W. Anderson, DSc, FInstBiol, FRSE; D. E. Butler, BSc, PhD and A. A. Morter, BSc
formerly of the British Geological Survey

Other publications of the Survey dealing with this district and adjoining districts

BOOKS

British Regional Geology
The Wealden District, 4th Edition, 1965

Memoirs
Tunbridge Wells (303), 1972
Tenterden (304), 1966
Brighton and Worthing (318/333), 1987
Hastings and Dungeness (320/321), 1987

Reports
Broadoak Borehole, Sussex, No. 78/3, 1978
Boreholes in the Wealden Beds of the Hailsham area, Sussex, No. 78/23, 1978

Well Catalogue
Records of Wells in the area of Lewes (319), Hastings (320) and Dungeness (321) sheets, 1965

MAPS

1:625 000
Solid geology of Great Britain (South)
Quaternary geology of Great Britain (South)
Sheet 2 Aeromagnetic

1:50 000 and 1:63 360 (Solid and Drift)
Sheet 302 (Horsham), 1972
Sheet 303 (Tunbridge Wells), 1971
Sheet 304 (Tenterden), 1981
Sheet 318/333 (Brighton and Worthing), 1984
Sheet 319 (Lewes), 1979
Sheet 320/321 (Hastings and Dungeness), 1980
Sheet 334 (Eastbourne), 1979

Printed in the United Kingdom for Her Majesty's Stationery Office

Dd.238932 C20 2/88 398/2 12521

CONTENTS

PLATES

TABLES

PREFACE

The Lewes district is included in the Old Series Geological Sheet 5, which was surveyed on the one-inch (1:63 360) scale by W. T. Aveline, H. W. Bristow, W. B. Dawkins, C. Le Neve Foster and W. Topley and published in 1864. The Chalk areas of the New Series Sheet 319 (Lewes) were re-surveyed on the six-inch (1:10 560) scale by W. A. E. Ussher and C. Reid in 1884 and 1890 and the river deposits were re-examined by F. H. Edmunds in 1925. The New Series Sheet 319 was issued in 1926 and the original Lewes memoir, written by H. J. O. White, was published in the same year.

The primary six-inch survey of the sheet was carried out by Dr C. R. Bristow and Messrs R. D. Lake and B. Young in 1970–73 under Mr S. C. A. Holmes and Dr W. A. Read as District Geologists. Small areas had been re-surveyed in 1964–65 and 1968 by Dr R. A. B. Bazley and Dr R. W. Gallois as overlap from the adjacent Horsham (302), Tunbridge Wells (303) and Hastings (320) sheets. The six-inch revision survey of the Chalk areas was conducted by Messrs Lake and Young in 1973. A list of six-inch maps and the names of the surveyors are given in Appendix 1. The 1:50 000 map of the district was published in 1979. Messrs Lake and Young have been responsible for compilation of the memoir, which was completed under Dr Bazley and Dr Read as District Geologists. This manuscript has been edited by Mr G. Bisson.

Palaeontological work on specimens collected from the district has been the responsibility of the Biostratigraphy Group of BGS, with assistance from other specialists. Micropalaeontological determinations on samples recovered from the Grove Hill (Hellingly) Borehole have been carried out by Dr G. Warrington, Professor W. G. Chaloner (Royal Holloway and Bedford New College), Dr A. R. Lord and Dr J. E. Robinson (University College, London); the Jurassic fauna of the Brightling and Grove Hill boreholes by Dr H. C. Ivimey-Cook and Dr B. M. Cox; the Wealden ostracods by the late Dr F. W. Anderson; the Lower Cretaceous charophytes by Dr M. Feist (Montpelier University, France); the Wealden bivalves and the Lower Greensand and Gault faunas by Dr R. Casey and Mr A. A. Morter; the micropalaeontology of the Gault and Lower Chalk by Professor M. B. Hart and his research students of Plymouth Polytechnic and by Dr D. J. Carter of Imperial College; the Chalk fauna by Mr C. J. Wood. Dr R. N. Mortimore of Brighton Polytechnic has provided a wealth of original stratigraphical information on the Chalk and this contribution is particularly acknowledged. The hydrogeology of the district has been described by Mr R. A. Monkhouse. The fuller's earth beds in the Hampden Park Borehole were identified by Mr R. J. Merriman, who also examined the mineralogy of certain Wealden clays. The most recent photographs were taken by Mr C. J. Jeffery. Several boreholes were drilled by the BGS drilling rig under the supervision of Mr S. P. Thorley.

I gratefully acknowledge the information and assistance afforded by officials of British Gypsum Ltd, the Eastbourne Waterworks Company, East Sussex County Council, and the Southern Water Authority, and the cooperation of landowners and quarry operators during the course of the survey. The British Petroleum Company kindly allowed the use of material from an unpublished report on the Pevensey area by the late Mr A. H. Taitt.

F. Geoffrey Larminie, OBE
Director

British Geological Survey
Keyworth
Nottinghamshire

9 December 1987

NOTES

The word 'district' is used in this memoir to mean the area represented by the 1:50 000 Geological Sheet 319 (Lewes).

National Grid references are given in square brackets throughout the memoir. They all lie within the 100 km square TQ (or 51) except where otherwise indicated.

Numbers preceded by A refer to photographs in the Survey's collections.

Numbers preceded by BGS, GSM or BGS GSM refer to specimens in the BGS biostratigraphical collection.

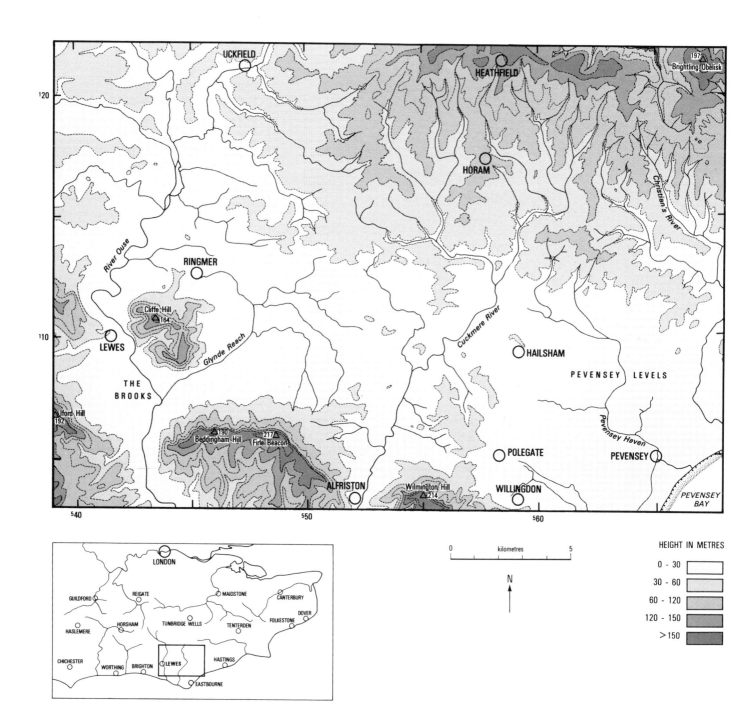

Figure 1 Physiography and drainage of the Lewes district

CHAPTER 1

Introduction

The country represented by the Lewes 1:50 000 geological sheet lies within the county of East Sussex and extends from the hills of the central Weald to the South Downs and the coast at Pevensey Bay (Figure 1). Lewes, the County Town of East Sussex, is near the western edge of the district, and other important centres of population are Uckfield and Heathfield in the extreme north, Hailsham, and Polegate and Willingdon in the south. The district is mainly agricultural, with many scattered villages and some large areas of forest. Apart from the cement works near Lewes and a few small engineering factories there is little industry. Parts of the district are popular as residential centres, with considerable numbers of people travelling to work in London each day, particularly from the Uckfield, Lewes and Polegate areas.

GEOLOGICAL SEQUENCE AND SCENERY

The formations that occur in this district are listed on the inside front cover. The Jurassic rocks were deposited on a surface of folded and faulted Palaeozoic rocks. The Brightling No. 1 Borehole [6725 2182] proved marine Devonian sediments below faulted Jurassic strata (Figure 3), whereas the Grove Hill Borehole [6008 1359] reached sediments possibly of Triassic age, not Carboniferous as was previously thought.

The district forms part of the southern limb of the Wealden dome, a large complex anticlinorium occupying much of the south-east of England. In the Lewes district the rocks at the surface have a gentle regional dip to the south-south-west, the oldest rocks being exposed in the north-east of the area and being succeeded by WNW–ESE-trending outcrops of progressively younger rocks (Figure 2). There are numerous minor structural modifications to this overall pattern and in the southern part of the district a more important structural element affects the rocks. Here, between Lewes and Polegate, the WNW–ESE-trending Kingston–Beddingham Anticline and Caburn Syncline disrupt the general south-south-westerly dip.

Within the district three broad belts of distinctive scenery may be recognised, each reflecting the local geology. The first of these belts corresponds to the outcrop of the Hastings Beds, together with the small inlier of Purbeck Beds in the extreme north-east. This belt is part of the 'High Weald', the area of undulating, hilly country occupying the central part of the Wealden dome. In the Lewes district the Hastings Beds reach 197 m above sea level near Brightling obelisk [6700 2121]. The ground slopes south and south-westwards from here, the Hastings Beds outcrop being characterised by rolling country with deeply incised, steep-sided, wooded valleys. The sandy formations of the Hastings Beds, the Ashdown Beds and Tunbridge Wells Sand, include a number of sandstones which weather to give slight benches along many hillslopes. These sandstones are exposed in numerous sunken lanes. They give rise to generally well-drained, light silty to sandy soils with large areas of woodland as well as good arable land. Spring-lines often occur at the junction of sandy beds with clays or silty clays. The Wadhurst Clay outcrop is marked by rolling rather featureless country, usually with a great number of small flooded pits from which clay or 'marl' was formerly dug for the treatment of sandy soils. Degraded bell-pits are common near the base of the formation and testify to the former importance of the Wealden iron industry. The wet, heavy clay soils on the Wadhurst Clay are mostly given over to pasture land.

The land on the Hastings Beds outcrop falls away gently to the second area of scenery, which is a broad belt of lowland marking the outcrop of the Weald Clay and overlying formations. The Weald Clay gives rise to slightly undulating country sloping gently south-south-westwards. Locally thin beds of limestone ('Sussex Marble'), ironstone and sand form minor scarps, though these harder beds are much fewer here than in other areas of the Weald. In the south-east of the district the broad alluvium-filled Pevensey Levels have been eroded into the Weald Clay, though an inlier of Tunbridge Wells Sand forms the higher ground on which Hankham stands. Much of the Weald Clay country and Pevensey Levels is pasture land. This median lowland area extends across the outcrops of the two succeeding formations, the Lower Greensand and the Gault; in contrast to the Lower Greensand outcrop of the western and northern Weald, the Lower Greensand here has a subdued topographic expression. Farther west, the Gault in the core of the Kingston–Beddingham Anticline has been eroded to give the wide alluvium-filled basin known as The Brooks (Plate 8).

The third belt of distinctive scenery, occupying the south-west corner of the sheet area, is part of the Chalk uplands of the South Downs. The most striking topographic feature of the district is the steep north-facing scarp of the South Downs (Plate 1). The scarp rises sharply from the lowlands to the north, though its foot seldom corresponds exactly to the base of the Chalk. The Lower Chalk usually forms the lower slopes of the escarpment as well as a tract of undulating, commonly heavily drift-covered ground, extending for some distance away from the scarp. The main slope of the escarpment and the dip-slope are on the Upper and Middle Chalk, and the Melbourn Rock at the base of the latter generally forms a prominent step. The highest points of the Chalk downland in the district are on the crests of the main escarpment; Firle Beacon [4857 0593], the highest point, reaches 217 m, Wilmington Hill [5484 0343] is 214 m high (Figure 1), and several other hills along the main escarpment exceed 180 m. The escarpment face is strongly embayed by coombes (Plate 7) which are believed to have been produced by nivation in glacial or periglacial conditions. Extensive solifluction lobes of chalky Head or Coombe Deposits, which blanket areas at the foot of the escarpment, originated in these coombes. Bull (1936) described a series of erosion surfaces preserved on the face of the escarpment. The higher areas of the Downs form a SSW-sloping dip-slope, dissected by numerous steep-sided, commonly sinuous dry valleys. Much of the downland was formerly used as pasture for sheep grazing, though today an increasingly large area has been ploughed and produces grain crops.

Around Lewes the configuration of the Chalk outcrop is complicated by the Caburn Syncline and Kingston–Beddingham Anticline, with Offham Hill, Cliffe Hill and Mount Caburn forming part of the first structure, and The Brooks and the valley of Glynde Reach at Beddingham on the latter (Plate 8).

DRAINAGE

The district is drained by streams which flow south to the English Channel (Figure 1). The two main river systems are the Ouse and the Cuckmere. The eastern part of the area is drained by shorter streams including the Pevensey Haven and the Waller's Haven which flow across Pevensey Levels to Pevensey Bay. A small area in the extreme north-east is drained by the River Dudwell, a tributary of the River Rother. A modified trellis pattern of tributaries is apparent in the Ouse and Cuckmere systems and to a lesser extent in the Waller's Haven drainage. The River Ouse downstream from Isfield is flanked by broad stretches of alluvium and a system of poorly-preserved terraces. At Lewes the Ouse flows through a narrow gap cut through the Chalk in the Mount Caburn synclinal structure, emerging to cross the flat vale of The Brooks south of Southerham before cutting through the Chalk once more in the wide valley at Southease. Whilst crossing The Brooks the Ouse receives a small left bank tributary, the Glynde Reach, which rises in the flat alluvial basin of Laughton Levels. In the valley known as Glynde Gap (Plate 8), between Beddingham and Glynde, the Glynde Reach is a striking example of a misfit stream. Kirkaldy and Bull (1940) suggested that the Glynde Gap was cut originally by a major tributary of the Ouse consisting of the present Uckfield branch of the Ouse flowing south to include the drainage basin of the present Glynde Reach; the Uckfield branch of the Ouse was claimed to have been captured by the stronger main Ouse stream near Isfield. The Cuckmere River has a narrow belt of alluvial country with terraces in places.

The courses of the Ouse and the Cuckmere have been artificially straightened in recent historical times, leaving a number of ox-bow lakes or cut-offs in the lower Ouse valley. Both rivers are known to have deep buried channels below their lower courses, and the base of the alluvial infill of Pevensey Levels also descends at least 15 m below present sea level.

GEOLOGICAL HISTORY

The Jurassic rocks were laid down in a fully marine environment in which shelf limestones and mudstones were deposited. Shephard-Thorn and others (1966) have given a brief account of the depositional history of these sediments in the eastern part of the Weald. In the Lewes district the oldest

rocks that crop out are the Purbeck Beds (Figure 2). These were deposited in a lagoonal environment resulting from the late Portlandian regression. The basal part of the Purbeck succession includes several beds of anhydrite and gypsum, reflecting a hot contemporary climate with relatively restricted circulation of water within the sedimentary basin. The Purbeck Beds overlying the evaporite beds consist of a rhythmic sequence of mudstones and limestones and provide evidence from their ostracod and bivalve faunas of conditions of widely varying salinity. One limestone horizon, however, that of the Cinder Bed, contains a fully marine fauna and represents a short-lived transgression; this has been taken to mark the base of the Cretaceous System (Casey, 1963), but recent evidence suggests that the Purbeck strata below the Cinder Bed may also be Cretaceous in age (see p.14).

The Purbeck Beds are overlain by the Wealden Beds, a sequence of clays, silts and fine-grained sands deposited in a continuing shallow lagoonal environment. In the older deposits the silts and sands are sufficiently abundant to comprise thick formational units, whereas in the Weald Clay only thin impersistent sand beds occur within the dominantly argillaceous sequence. Allen (1954; 1959; 1962; 1967a; 1975) has made detailed studies of the Wealden sediments and ascribed the influxes of coarse sediments to deltaic advances and latterly to invasion of a mudplain by braided or meandering river systems which locally deposited coalescent alluvial fans. These models placed the main sedimentary source in an upland region which is now the London area. However other authors (Kaye, 1966; Lake and Young, 1978) have suggested a dominantly southerly source for the sediments. The influxes of sandy material may have been generated by movements of marginal fault blocks in the source area.

A broad cyclicity of sediments may be recognised both at formational level and on a much smaller scale in the more argillaceous sequences. This cyclicity resulted in some cases from transgression and regression of the lagoonal waters. The water within the basin was often shallow, and soil and rootlet beds are common at many levels. Hence a delicate balance existed between water-depth and sedimentation.

There is some evidence that a phase of minor warping and possible uplift and slight erosion of the Wealden Beds preceded the marine transgression which submerged the entire area and led to the deposition of the Lower Greensand and succeeding marine Cretaceous formations.

Occasional periods of still-stand and, at times, erosion occurred during the deposition of the Gault clays and the lowest bed of the Chalk, the Glauconitic Marl. By Cenomanian times, when the Lower Chalk was being deposited, a major marine transgression had submerged the whole of southern and eastern England and large areas of Europe. Successively higher parts of the Chalk, comprising debris of marine micro-organisms with little land-derived material, were deposited in wide, relatively shallow, warm seas. Minor erosion surfaces and hardgrounds record pauses in sedimentation and possibly episodes of emergence.

Phases of folding that produced the structural elements seen today occurred in late Cretaceous times and latterly in the Oligo-Miocene period. There were minor earth-movements throughout Cretaceous times which influenced

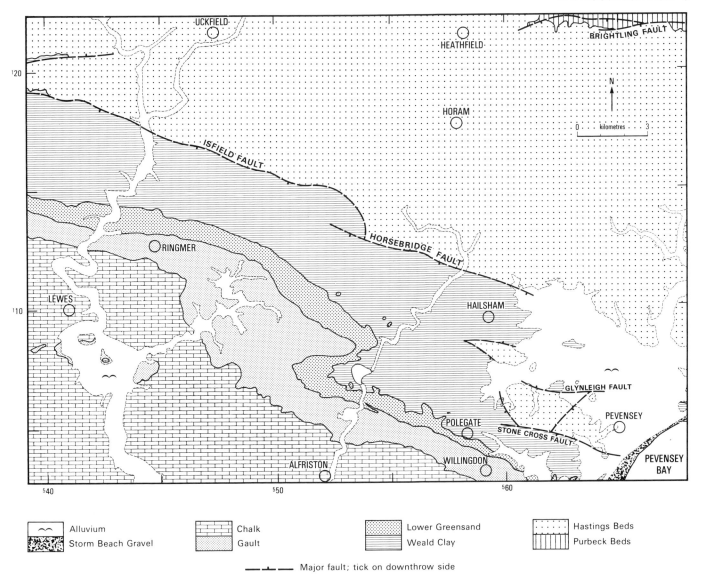

Figure 2 Geological sketch-map of the Lewes district

the pattern of sedimentation. For example, some of the divisions of the Chalk show significant variations both of thickness and lithology which resulted from the development of basins and swells. Although it is not possible to relate these depositional features directly to structural lineaments it is probable that they are genetically closely linked. The later earth movements transformed the Wealden Basin, a subsiding region in which the Mesozoic sediments accumulated, into an area of uplift which shows a broad anticlinorial structure in the upper part of the sedimentary pile. During the Palaeogene, the deposition of clastic sediments was apparently centred over more restricted sedimentary basins but possibly extended over much of the Weald although confirmatory evidence is lacking. A further period of partial submergence during the Pliocene period may have permitted the deposition of marine sediments over parts of the Weald, though much of the central Weald probably remained as an island at this time. The early Tertiary and Pliocene deposits, almost certainly laid down over large parts of the Lewes

district, have been entirely removed by erosion in late Pliocene and Quaternary times, although outliers of Tertiary deposits remain at Newhaven and Falmer, respectively short distances south and west of the present district.

During the Quaternary the Weald suffered extensive subaerial erosion which produced the present-day drainage pattern and landscape. In the later part of the Pleistocene period the area was at times subjected to periglacial conditions which gave rise to locally extensive solifluction deposits. The deep buried valleys of the Ouse, Cuckmere and parts of Pevensey Levels record a period of lowered sea level, but the progressive rise of sea level in Flandrian times has resulted in the silting up of these channels.

PREVIOUS RESEARCH

Lewes was the home of Gideon Mantell, best remembered for his discovery of the dinosaur *Iguanodon* and also for his

Table 1 British Geological Survey cored boreholes

Borehole	National Grid reference	Stratigraphical range	Reference
Alciston	5045 0553	Lower Chalk – Gault	in text
Broadoak	6195 2214	Purbeck Beds – Portland Beds	Lake and Holliday, 1978
Cooden	7043 0641	Weald Clay – Ashdown Beds	Lake, 1975a
Cuckfield No. 1	2962 2729	Weald Clay – Ashdown Beds	Lake and Thurrell, 1974
East Hoathly	5186 1603	Tunbridge Wells Sand – Ashdown Beds	Lake and Young, 1978
Glyndebourne	4420 1141	Lower Chalk – Lower Greensand	in text
Glynleigh	6085 0637	Tunbridge Wells Sand – Ashdown Beds	Lake and Young, 1978
Hailsham	5746 1083	Weald Clay – Tunbridge Wells Sand	Lake and Young, 1978
Hampden Park	6120 0204	Gault – Weald Clay	Lake and Young, 1978* Young and others, 1978
Ripe	5059 1052	Lower Greensand – Tunbridge Wells Sand	Lake and Young, 1978*
Rodmill	6008 0070	Lower Chalk – Gault	in text
Warlingham	3476 5719	Mesozoic – Carboniferous	Worssam and Ivimey-Cook, 1971
Westfield	8204 1614	Tunbridge Wells Sand – Wadhurst Clay	Lake and Shephard-Thorn (in preparation)
West Firle	4676 0809	Lower Chalk – Gault	in text

* Wealden beds only

work in describing Wealden geology and establishing the freshwater origin of the Wealden Beds (1818; 1827; 1833). Much of the early geological literature dealing with the Weald is relevant to the Lewes district, in particular the work of Drew (1861). Topley (1875) produced the first comprehensive account of Wealden geology.

The Purbeck strata of Sussex have been the subject of works by Casey (1963), Howitt (1964), Norris (1969), Anderson and Bazley (1971), Holliday and Shephard-Thorn (1974), and Lake and Holliday (1978).

Considerable progress towards a detailed understanding of the Wealden environment has been made by Allen (e.g. 1941; 1954; 1959; 1960a; 1960b; 1961; 1962; 1967a; 1967b; 1975; 1981) and the Weald Clay has been investigated by Reeves (1949; 1958) and Lake and Young (1978). Notable contributions to our knowledge of the Lower Greensand and Gault have been made by Jukes-Browne and Hill (1900), Kirkaldy (1935; 1937) and Casey (1950; 1960b; 1961).

Several authors have dealt with the Chalk, some of the earliest descriptions of the local Chalk being given by Mantell (e.g. 1822) and Dixon (1850). The studies of the English Chalk by Barrois (1876) and Rowe (1900) include references to the Lewes district. Other important works are by Jukes-Browne and Hill (1903; 1904), Gaster (1929; 1937a; 1939; 1951) and Kennedy (1967; 1969), and Gaster's work in particular has provided a wealth of detail on the

zonal classification of the Chalk of the Lewes district. Mortimore (1979; 1986a) has made a detailed study of the Chalk of Sussex and his work has been incorporated in this account. Aspects of the sedimentology of the Chalk have been discussed by Bromley (1967; 1978), Hancock (1975), Hancock and Kennedy (1967), Hofker (1959), Lombard (1956), Voigt (1959) and Wood (*in* Worssam and Taylor, 1969). The origin of flint has been treated by Shepherd (1972), Bromley and others (1975), and Hancock (1975).

The geomorphology of the Weald has been the subject of work by Wooldridge and Linton (1955), and references to the geomorphology of the Lewes area are included in papers by Bull (1936), Kirkaldy and Bull (1940), Williams (1971) and Jones (1971). The area depicted by the Lewes (319) Sheet includes the site of the famous Piltdown skull hoax (Dawson and Woodward, 1913).

Straker (1931) has made a detailed study of the Wealden iron industry and a number of important sites are in the Lewes district.

A Geological Survey Catalogue of wells in the Lewes district has been published (Cole and others, 1965). Information from cored boreholes drilled by BGS in East Sussex and elsewhere in the Weald has enabled the stratigraphy of the region to be refined and new correlations to be made. The BGS boreholes in Table 1 are referred to in the text (see also Appendix 2).

RDL, BY

CHAPTER 2

Concealed formations

INTRODUCTION

Two boreholes have provided core samples from strata older than the Purbeck Beds in the Lewes district, namely the Grove Hill Borehole (sometimes referred to as the Hellingly Borehole), drilled by the Anglo-American Oil Company Limited in 1937, and the Brightling (or Coombe Hill) Borehole (more correctly termed the Brightling No. 1 Borehole), drilled by the British Petroleum Company Limited in 1955 and extended in 1957. Specimens from these boreholes have been presented to BGS by the companies and their interpretation has been facilitated by the logs made at the time of drilling by geologists of the companies, and by the late Mr A. Templeman of the Geological Survey. The specimens have been re-examined by Dr B. M. Cox (Upper Jurassic), Dr H. C. Ivimey-Cook (Middle and Lower Jurassic and adjacent rocks), and by Dr D. E. Butler (Devonian). Dr G. Warrington has reported on the Lower Jurassic microflora from Grove Hill, and we are grateful to Professor W. G. Chaloner for his comments on the early Jurassic flora and to Dr J. E. Robinson and Dr A. R. Lord for their comments on the early Jurassic ostracods. Little has been published on these boreholes except in the review by Falcon and Kent (1960). The strata penetrated in these boreholes are summarised below and are shown in Figure 3.

In addition, the Westham Borehole [6097 0535] was drilled by Cambrian Exploration Limited in 1973. The summarised record below is based largely on examination of rock chippings and electric logs. The depths are related to the rotary table at 8.5 m OD, 5.2 m above ground level. The formations above the Corallian are generally thinner than they are at Grove Hill and Brightling. In particular the Hastings Beds (i.e. the beds below the middle part of the Tunbridge Wells Sand) and the Purbeck Beds show a marked attenuation, very similar to that seen in the Pevensey No. 1 Borehole (p.14). RDL

Grove Hill TQ 6008 1359	Thickness m	Depth to base* m
Purbeck Beds	about 103.6	310.9
Portland Beds	24.4	335.3
Kimmeridge Clay	286.5	621.8
Corallian Beds	76.2	698.0
Oxford Clay	76.2	774.2
Kellaways Beds	6.7	780.9
Great Oolite Group	71.6	852.5
Inferior Oolite Group	35.4	887.9
Lias	about 166.7	about 1054.6
?Triassic	seen to 14.0	to final depth of 1068.6

* Original measurements were expressed in feet, and the figures obtained on converting these to metres have been rounded off to one place of decimals.

Brightling TQ 6725 2182	Thickness m	Depth to base* m
Purbeck Beds	about 112.8 +	112.8
Portland Beds	31.7	144.5
Kimmeridge Clay	407.2	551.7
Corallian Beds	76.8	628.5
Oxford Clay	68.9	697.4
Kellaways Beds	7.9	705.3
Great Oolite Group	69.8	775.1
Inferior Oolite Group	80.5	855.6
Lias	about 282.0†	1322.2
Devonian (probably Upper)	seen to ?183.5‡	to final depth of 1505.7

* Original measurements were expressed in feet, and the figures obtained on converting these to metres have been rounded off to one place of decimals.

† The thickness given for the Lias is that of the sequence above a major fault; about two-thirds of this succession is seen again below the fault (see pp.7 – 10).

‡ Only the rocks between 1502 and 1505 m, close to the final depth of this borehole, have yielded an Upper Devonian fauna.

Westham	Thickness m	Depth m
Hastings Beds (Tunbridge Wells Sand 45 m*; Wadhurst Clay 30 m*; Ashdown Beds 111 m*)	185.9	185.9
Purbeck Beds	53.4	239.3
Portland Beds	27.4	266.7
Kimmeridge Clay	205.7	472.4
Corallian Beds	?93.6	?566.0
Oxford Clay, including Kellaways Beds	?89.3	655.3
Great Oolite Group, including 22 m of Fuller's Earth clays at base	67.7	723.0
Inferior Oolite Group	30.5	753.5
Lias and late Triassic (Penarth Group)	307.8	1061.3
Lower Carboniferous (?faulted)	176.2	1237.5
Upper Devonian (Frasnian/Famennian)	seen to 360.3	1597.8

* Figures derived from the record of the Glynleigh Borehole, p.103.

PALAEOZOIC

In the Brightling Borehole Palaeozoic rocks are thought to have been entered at about 1322.2 m. Core taken just below this depth yielded purple, brown and green mudstone with a dip of 60° decreasing rapidly downwards to an average of 25° to 30°. The mudstone is shattered and slickensided, giving support to the suggestion (p.7) that the contact with the Lias above is faulted. Further cores were taken at intervals to 1391.4 m, where hard calcareous sandstones, contorted shaly mudstone and grey siltstone occur. A terminal core was taken between 1502.6 and 1505.7 m and only this yielded faunal evidence of the age of the sequence. Early work on the macrofossils (Falcon and Kent, 1960, p.11) and on the spores (Mortimer and Chaloner, 1972) from this core sug-

gested a Lower Devonian age but re-examination of the macrofossils (specimens BGS Zi7995–8033A) from 1502.6 to 1505.4 m by Butler (1981) led him to refer them to the Upper Devonian. He recorded the brachiopods *Lingula punctata*, *Cyrtospirifer?*, and *Retichonetes sp. nov.*, the bivalves *Leptodesma spinigerum*, '*Nuculana*' sp. nov., *Nuculoidea corbuliformis*, *Palaeoneilo constricta*, *P.* aff. *maxima*, *Paracyclas tenuis* and *Pseudaviculopecten* aff. *striatus*, and a tentaculitoidean *Dicricoconus sp. nov.* The five established species range up to or through the Frasnian from earlier horizons, whereas the three new species are known from elsewhere in the British subsurface Upper Devonian. *Pseudaviculopecten striatus* itself is known from the Frasnian in North America. Thus a Frasnian age was favoured for this fauna, although the evidence was not regarded as conclusive.
DEB

No later Palaeozoic rocks were proved in either the Brightling Borehole or the Grove Hill Borehole. Re-examination of the specimens available from the Grove Hill Borehole has cast doubt on it having entered the Palaeozoic at all. The material from between 1051.6 and 1054.0 m, which had previously been considered to be Carboniferous, has been found to contain an early Jurassic microflora (below). Beneath these beds are about 1 m of fissured off-white calcilutites, together with some chert and mottled marls, from which no evidence of age has been obtained. These limestones have a slightly nodular fabric with a mesh of clayey matter. The overall appearance is similar to that of an anhydrite rock with a penemosaic texture and thus a secondary origin seems likely (Dr D. B. Smith, personal communication). The borehole apparently continued for a further 9 m (reported as 'marl'), from which no specimens appear to have been kept. Lithologically the calcilutites do not closely resemble Palaeozoic limestones proved in nearby boreholes and although their age must be regarded as unproven it is possible that they are of Triassic age and that the Mesozoic was not completely penetrated at Grove Hill.
HCI-C

In the Westham Borehole brown shales and micritic limestones alternate below 1237.5 m depth. Dr B. Owens reports that the miospore assemblages of samples taken between 1246.6 and 1496.6 m indicate an Upper Devonian age. Grey shales, siltstones and siliceous limestones occur between 1061.3 and 1237.5 m; the upper limit of this division was taken arbitrarily at the top of a bed of purplish red chert, although superficially similar but less siliceous limestones occur above. The miospore assemblages of samples taken between 1155.2 and 1161.3 m and between 1213.1 and 1219.2 m are reported by Dr Owens to indicate a Tournaisian age, but Carboniferous miospores found in association with bisaccate pollen of Mesozoic aspect at 1103.4 m are possibly reworked. Dr G. Warrington and Mr J. B. Riding recovered palynomorph assemblages indicative of Jurassic ages no younger than early Bajocian at 868.7 m; they found no evidence for a pre-Jurassic age in palynomorph assemblages recovered down to 1118.6 m. The dating of the sequence between 868.7 and 1155.2 m is not supported by biostratigraphic evidence. Similarly the presence of reverse faulting in this part of the succession, as suggested in the original log on the basis of apparently duplicated lithological sequences above and below 1192 m, has not been confirmed.
RDL

LOWER JURASSIC

The Lower Jurassic rocks in the Brightling Borehole are largely repeated below a fault at 1137.5 m. The upper sequence, between 855.6 and 1137.5 m, is the thicker (281.9 m) and it contains Upper, Middle and Lower Lias; eight short cores were taken. The lower sequence, between 1137.5 and 1322.2 m, has less of both the Upper Lias and the Lower Lias but 184.7 m of beds are present; much of the succession was cored except for the lowest 42 m. In the Grove Hill Borehole the Lower Jurassic was about 167 m thick, and very little now exists of the six short cores taken (see Figure 3).

Lower Lias

In the Grove Hill Borehole the Lower Lias is between 55 and 89 m thick, whereas at Brightling it is of the order of 158 m. The lithologies are similar (see Figure 3). At Grove Hill the lowest beds are dull greyish green and dark brownish red marl with plant remains and hard grey and purplish brown limestone with mottled siltstone occur between the depths of 1051.6 and 1054.0 m. The macrofauna contains bivalves including '*Gervillia*' sp. (cf. *Pteria alfredi*), *Liostrea* sp. and *Modiolus minimus?*, and small gastropods including *Coelostylina* aff. *nana* and a cerithiid. A near-shore environment is suggested by the presence of small smooth-valved ostracods now found as external moulds. This fauna is considered to be early Jurassic in age.

A macroflora, including the remains of a large compound pinnate leaf (cf. *Ctenis*) identified by Professor Chaloner, occurred between 1051.6 and 1052.2 m. From the same depth range Dr Warrington has reported as follows on the microflora: 'A moderately well preserved and fairly profuse and varied palynomorph assemblage comprising miospores, organic-walled microplankton and the tectinous test linings of foraminifera; organic debris, including plant tissues of medium brown colour, is also present.'

'Miospores are the dominant component of this assemblage (93 per cent based upon a count of 200 palynomorphs) and include the following: *Alisporites thomasii*, *Chasmatosporites apertus*, *Classopollis torosus*, *Cyathidites minor*, *Kraeuselisporites reissingeri*, *Leptolepidites argenteaeformis*, *Perinopollenites elatoides* and *Tsugaepollenites mesozoicus*. The miospore association is dominated by bisaccates and *Classopollis torosus* which form, respectively, 46 per cent and 40 per cent of the total palynomorph assemblage. The remaining taxa are mostly represented by single specimens, exceptions being *K. reissingeri* and *T. mesozoicus* which both constitute 2.5 per cent of the assemblage.'

'Acanthomorph acritarchs dominate the organic-walled microplankton component of the assemblage which includes the following: *Crassosphaera hexagonalis*, *Micrhystridium fragile*, *M. lymense* var. *gliscum*, *M. lymense* var. *rigidum*, *M.* cf. *rarispinum* and an indeterminate tasmanitid.'

'The palynomorph assemblage is indicative of an horizon in or equivalent to the lower part of the Lias sequence and its age, on the basis of the association of *K. reissingeri* with *T. mesozoicus*, is probably not older than early Sinemurian. The presence of organic-walled microplankton and foraminifera is indicative of deposition in a marine environment. The preponderance of miospores in the assemblages also suggests that the site may have been close to land.'

Core taken between 1041.8 and 1043.9 m contains grey sandy limestones with echinoderm, juvenile ammonite, *Pseudolimea?*, inoceramid and ostreid fragments and a belemnite, which are thought to be of Sinemurian age. Core between 1002.8 and 1005.2 m contains silty mudstone with *Palaeoneilo?* which is not diagnostic of age. The Lower Lias may continue up to between 1002 and 968 m but little information is available. The core between 964.7 and 967.7 m is grey slightly pyritic mudstone. It could be from either the Lower or Middle Lias.

In the Brightling Borehole the Lias sequence contains evidence of several faults (Figure 3). A major reversed fault, which duplicates almost all the Lias sequence, is present at about 1137.5 m. In addition a fault is postulated at the Lias to Palaeozoic junction at 1322.2 m, where the original log records (it was not cored) veined pale gritty pyritic sandstone on shattered purple, brown and green (Palaeozoic) mudstone with high and variable dips and much veining. The lower sequence of Lias also appears to lack beds present near the base of the upper sequence. Other faults are suspected within the Lias but are difficult to prove. Cores were taken at intervals in the upper sequence and considerably more core in the lower but many details of the zonation remain speculative.

At Brightling the Lower Lias is estimated to be about 158 m thick, occurring both between about 975.4 and 1137.5 m above the fault and between 1197.9 and 1322.2 m below it. This thickness allows for the recorded dips of about 15° above and 5° below the fault. However, there may also be some repetition of beds in the *Prodactylioceras davoei* Zone and other beds may also have been cut out by faults.

In the upper sequence, fractured grey silty limestone and calcareous mudstone immediately overlie the fault; they pass up into less calcareous mudstones and then into mudstones with subordinate limestones. A dark reddish brown oolitic ironstone is present at 1071.7 m. The oldest age proved is the *Asteroceras obtusum* Zone, about 60 m above the fault and it is probable that beds of Lower Sinemurian age are also present in this thickness. However the core between 1123.19 and 1129.28 m is of grey brecciated limestone and has not yielded diagnostic fossils.

Promicroceras sp., found between 1077.7 and 1084.8 m, occurs in the higher part of the *Caenisites turneri* Zone and throughout the *obtusum* Zone. The presence of the latter zone is indicated by *Galaticeras sp.* at 1077.8 m. *Piarorhynchia sp.* and bivalves are also present. No evidence for the *Oxynoticeras oxynotum* Zone is known. In the *Echioceras raricostatum* Zone the *Crucilobiceras densinodulum* Subzone is indicated by *Bifericeras*, *Crucilobiceras* and *Hemimicroceras* between 1268.1 and 1271.8 m, whereas in the upper sequence the middle of this zone is indicated by *Paltechioceras boehmi*. The higher part of the *raricostatum* Zone and the overlying *Uptonia jamesoni* Zone are unproved but, with the *Tragophylloceras ibex* Zone indicated at 1251.2 m, the combined thickness of the *raricostatum* and *jamesoni* zones may not exceed 20 m unless affected by faulting. The *ibex* Zone also appears to be very thin, with the *Acanthopleuroceras valdani* Subzone ammonite *Acanthopleuroceras?* at 1251.2 m and *Aegoceras (Beaniceras) sp.*,

probably of *Beaniceras luridum* Subzone age, between 1247.4 and 1248.8 m. The zone could possibly extend up to the lowest *Androgynoceras* at 1246.0 m. Similarly the *A. maculatum* Subzone — with *Androgynoceras* cf. *maculatum* at 1242.6 to 1243.0 m — and the *A. capricornus* Subzone above are thin, as the lowest *Aegoceras (Oistoceras) sp.* is at 1238.25 m. In contrast, the *Oistoceras figulinum* Subzone appears to be unusually thick. *Oistoceras* is found in the lower sequence between 1230.3 and 1238.25 m, but no ammonites are known from the overlying core below the amaltheiid, indicating the *Amaltheus stokesi* Subzone of the *Amaltheus margaritatus* Zone, around 1189 m. In the upper sequence *A. (Oistoceras) figulinum* occurred at 1016.0 m and *A. (Oistoceras) sp.* between 1016.0 and 1019.5 m and also at 979.3 m. The similarity between these two cores suggests that there may have been repetition by faulting. The thickness of sediment deposited during the *figulinum* Subzone appears to have been significantly greater than that in earlier subzones. Similar lithologies persist into the Middle Lias but the junction has not been accurately located.

Middle Lias

In the Grove Hill Borehole the Middle Lias may occur between 967 and 934.5 m but very little information is available about these strata. The only specimens available are from between 934 and 936 m and consist of grey, calcareous, micaceous siltstone with possible fish fragments which are not diagnostic of age.

In the Brightling Borehole dark grey micaceous mudstones and siltstones with nodules of muddy limestone and scarce bivalves make up much of the Middle Lias; they include beds of *stokesi* Subzone age around 1189 m, where *Amaltheus* cf. *stokesi* is found. The mudstones become silty upwards and are succeeded by greenish grey ferruginous oolitic and shelly limestone and sandstone, possibly contemporaneous with the Marlstone Rock Bed: the age of this facies is not proved but it is overlain at 1149.7 m by beds of early Toarcian age.

Upper Lias

In the Grove Hill Borehole Upper Lias probably occurs above about 934 m. Above 896.4 m (the base of a cored sequence) grey micaceous siltstones and mudstones with a few shell fragments are overlain at about 894.3 m by grey limestones and calcarenites some of which are ferruginous, and, above 892.2 m, by brown ferruginous oolites and calcarenites cored below 887 m. These oolites are similar to beds proved to be of Upper Toarcian (*Dumortieria levesquei* Zone) age at Penshurst (Ivimey-Cook *in* Dines and others, 1969) and are included in the Upper Lias here.

In the Brightling Borehole early Toarcian mudstones with *Dactylioceras* occur between the top of the Middle Lias at 1149.7 m and the fault at 1137.5 m. In the upper sequence above this fault the Upper Lias dark grey mudstones are much thicker (41.2 m). A core taken at 910.9 m yielded a fragment of *Hildoceras bifrons*, indicating the *Peronoceras fibulatum* Subzone of the *bifrons* Zone of the Toarcian at an horizon 15.7 m above the base of the Upper Lias. These

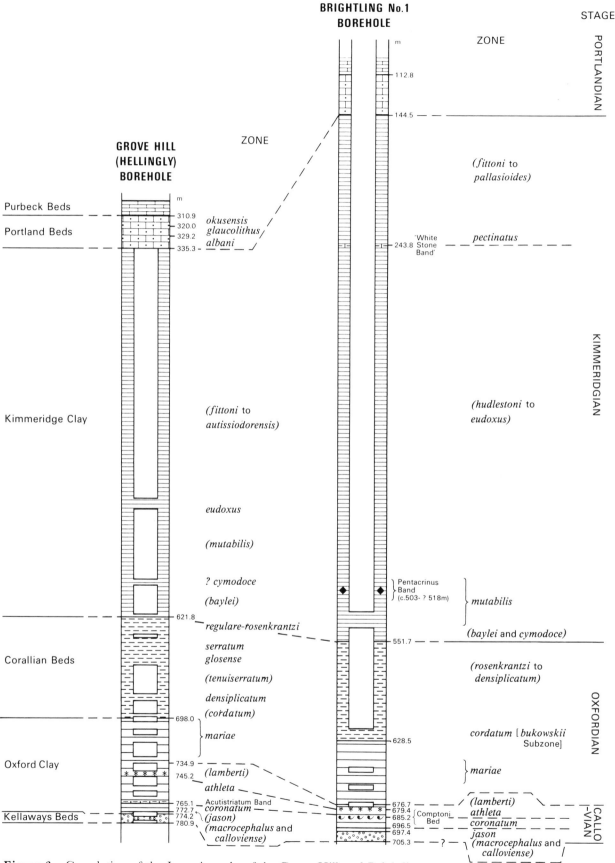

Figure 3   Correlation of the Jurassic rocks of the Grove Hill and Brightling boreholes

9

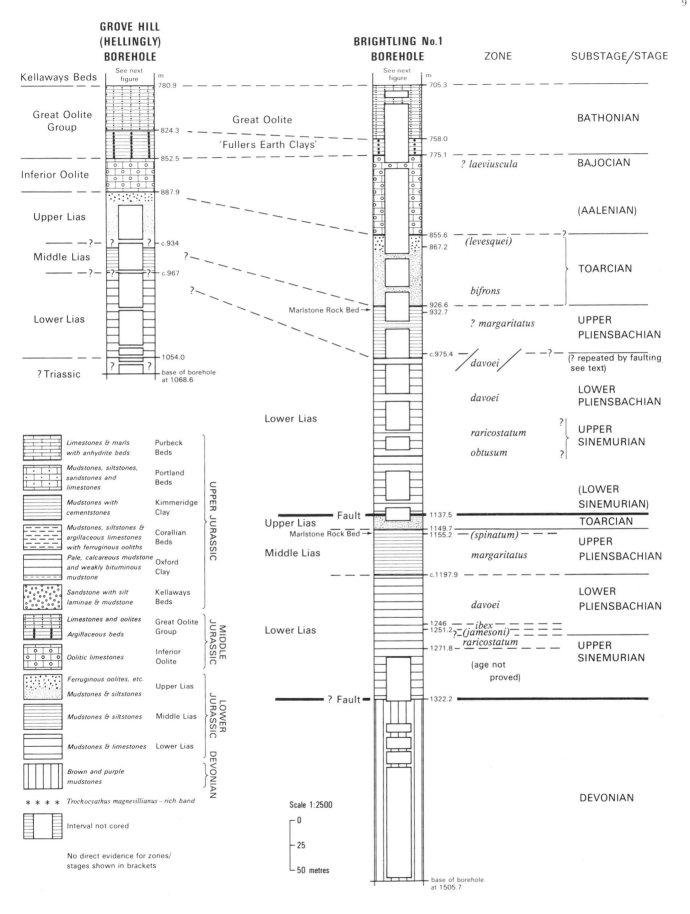

mudstones continue up to about 885 m and then become sandier. At 877.8 m a grey limestone with abraded and coated calcareous siltstone pebbles indicates a non-sequence but no faunal evidence of its age is available. Above, the log records that calcareous and micaceous siltstones and fine-grained sandstones are overlain by ferruginous oolitic limestones between 855.5 and 867.2 m. The latter can be compared with either the ferruginous beds of *levesquei* Zone age or the earlier ferruginous unit, of *Grammoceras thouarsense* Zone age, in the Penshurst Borehole (Ivimey-Cook *in* Dines and others, 1969) but no specimens are available.

The total thickness of Lias at Grove Hill is about 167 m, whereas at Brightling the Lias is about 282 m thick.

MIDDLE JURASSIC

Inferior Oolite Group

In the Grove Hill Borehole the sequence of crystalline and oolitic limestones between 852.5 and 887.9 m are attributed to the Inferior Oolite. A lower unit (about 23 m thick) of rather soft oolitic limestone with *Acanthothiris spinosa*, terebratuloids, *Chlamys sp.* and *Propeamussium sp.* is overlain by a higher unit of hard grey crystalline limestones with patches of ooliths. This higher unit yielded echinoderm fragments, brachiopods and bivalves including *A. spinosa* (853 to 861 m), *Stiphrothyris tumida* (852 to 853 m), *Camptonectes*, *Chlamys?*, *Entolium*, *Lopha?* and *Plagiostoma?*. These suggest it is part of the Upper Inferior Oolite.

From the Brightling Borehole very little core is available. The lithologies between 775.1 and 855.6 m include grey and fawn oolitic limestones with some crystalline and detrital limestone. Core from 782.1 to 788.2 m contains shelly pelletoid limestones with a macrofauna including *Sarcinella plexus*, *A. spinosa* and fragments of other brachiopods and bivalves. A small piece of an ammonite found at 782.12 m could be part of the whorl of a large sonniniid (cf. *Euhoploceras*) from the *Witchellia laeviuscula* Zone, in which case the upper part of these oolitic limestones would equate with the Middle Inferior Oolite as seen in the Henfield Borehole.

Great Oolite Group

In the Grove Hill Borehole the Great Oolite Group was extensively sampled. Calcareous mudstones and silty limestones, generally non-oolitic, form a lower argillaceous unit comparable with the Fuller's Earth Clay. An argillaceous limestone between 847.9 and 849.5 m contains *Rugitela cadomensis* and *Rhynchonelloidella*, and could be about the horizon of the Fuller's Earth Rock of the Bath area. Calcareous mudstones above and below this limestone contain a sparse bivalve fauna. Close above this limestone, at 847.3 m, *Acanthothiris?*, *Rhynchonelloidella smithi* and *Rugitela cadomensis* were recovered.

The Great Oolite itself comprises fine oolitic limestones with shell fragments and marl partings and a brachiopod-bivalve fauna including *Kallirhynchia*, *Lucina?*, *Lopha*, *Modiolus?*, *Propeamussium* and *Protocardia?* between 791.3 and 824.3 m.

Late Bathonian 'Forest Marble' sediments comprise alternating argillaceous limestones with scattered ooliths, oolitic limestones, calcarenites and calcareous mudstones from 781.5 to at least 788 m. Their macrofauna includes echinoderm fragments, *Digonella* cf. *digonoides*, *Rhactorhynchia*, *Kallirhynchia*, *Camptonectes*, *Entolium?*, *Liostrea*, *Modiolus*, *Palaeonucula*, *Placunopsis* and *Pseudolimea*; this fauna is comparable with that found near the base of the Forest Marble (Penn and others, 1979).

There is no sign of Cornbrash in the core samples and the 'Forest Marble' appears to be directly overlain by Kellaways Beds sandstones.

In the Brightling Borehole about 17.0 m of grey calcareous mudstones with some pale grey limestone form the lower part of the Great Oolite Group and are overlain by 52.7 m of pale grey oolitic and cream argillaceous limestone. Core from the upper part of this sequence (705 to 707 m) shows numerous coated pelloids set in a fine lime mud matrix with some shell fragments. No macrofossils were identified. HCI-C

UPPER JURASSIC

Kellaways Beds

The Kellaways Beds are best known from Brightling, where they consist of 7.9 m (697.4 to 705.3 m) of pale and brownish grey, bioturbated, fine-grained sandstone with silty laminae and scattered shells and shell fragments including *Meleagrinella*, *Oxytoma* and belemnites. The sandstone is calcite-cemented in part and there is some pyritisation. Similar lithologies, with *Pleuromya alduini* and *Catinula?*, are also present at Grove Hill, where a total thickness of about 6.7 m includes some uncored mudstone.

Oxford Clay

The Oxford Clay is traditionally divided into Lower, Middle and Upper on the basis of gross lithology and fauna (Callomon, 1968); all three divisions can be recognised in the boreholes illustrated in Figure 3.

The Lower Oxford Clay consists of grey, brownish grey and grey-green, fissile, faintly bituminous, shelly mudstones with some paler more calcareous beds. It is 18.3 m thick at Grove Hill (755.9 to 774.2 m) and 14.9 m thick at Brightling (682.5 to 697.4 m). The fauna is abundant and dominated by bivalves (Duff, 1978) including *Bositra buchii*, *Chlamys sp.*, *Corbulomima?*, *Discomiltha sp.*, *Grammatodon sp.*, *Isocyprina roederi*, *Meleagrinella braamburiensis*, *Neocrassina (Pressastarte) sp.*, *Nicaniella (Trautscholdia) phillis*, nuculoids including *Mesosaccella morrisi*, *Nuculoma sp.* and *Palaeonucula* cf. *calliope*, oysters, *Parainoceramus sp.*, *Pinna* cf. *mitis*, *Protocardia sp.* and *Thracia depressa*. *Bositra*, *Meleagrinella* and *Mesosaccella* are particularly common and occur in shell plasters and shell beds. Other fauna comprises the gastropods *Dicroloma sp.* and *Procerithium damonis*, the brachiopod *Discinisca sp.*, the serpulid *Genicularia vertebralis* (characteristic of the Lower–Middle Oxford Clay), and a belemnite fragment.

The ammonite fauna, which is partially pyritised and commonly has aragonitic shell material preserved, is dominated by *Kosmoceras* which indicate the presence of the three Callovian ammonite zones of *Kosmoceras jason*, *Erymnoceras coronatum* and *Peltoceras athleta*. The ammonites from the Grove Hill Borehole are: *Kosmoceras spp.* (including a looped-ribbed form) at 762.6, 764.4 and 764.7 m (*athleta* Zone), *Hecticoceras sp.* at 762.6, 763.2, 763.5 and 764.4 m, *Binatisphinctes sp.* at 763.2 m, *B. comptoni* var. *parvus* at 763.8 m, *Kosmoceras (Spinikosmokeras) acutistriatum* at 766.3 m(?) (*athleta* Zone), *K.(S.)* cf. *castor* at 765.1 and 766.6 m, *K. (S.)* aff. *castor* and *K.(S.)* cf. *herakles* at 766.3 to 766.9 m (*coronatum* Zone), *Erymnoceras sp.* at 766.3 and 767.8 m (*coronatum* Zone), *Hecticoceras sp.* at 766.6 m, *Kosmoceras spp.* at 767.2, 767.5 and 768.4 m, *Binatisphinctes?* at 767.5 m and perisphinctid fragments at 767.8 and 768.1 m, *Kosmoceras (Zugokosmokeras) obductum* at 767.8, 769.0, 769.3, 771.1 to 772.7, and 773.0 m (*coronatum* Zone), *Kosmoceras (Gulielmiceras) gulielmi* at 769.0, 769.6, 771.1 to 772.7, and 773.0 m, *K. (Spinikosmokeras) sp.* at 772.7 m and *Hecticoceras sp.* at 769.6 and 772.7 m; and those from the Brightling Borehole are: *Hecticoceras sp.* at 684.1, 684.2, 684.9 and 685.0 m, *Kosmoceras (Spinikosmokeras) sp.* at 684.2 m, *Binatisphinctes comptoni* at 685.2 and 685.3 m, *Hecticoceras sp.* at 685.5 m, *Kosmoceras (Spinikosmokeras) sp.* at 687.3, 695.6 and 695.9 m, *K. (Zugokosmokeras)* cf. *obductum* at 692.5 m (*coronatum* Zone), *Erymnoceras?* at 694.3 m (*coronatum* Zone) and *K. (Gulielmites)* cf. *jason* at 696.8 and 697.1 m (*jason* Zone). \

Throughout southern and central England, a pyritic nuculoid shell bed with *Binatisphinctes comptoni* (the Comptoni Bed) occurs at the top of the *coronatum* Zone and a pale calcareous mudstone or limestone bed with *Kosmoceras acutistriatum* and *B. comptoni* (the Acutistriatum Band) marks the base of the overlying *athleta* Zone (Callomon, 1968). At any one locality, these two marker horizons may not be equally well developed, but either can be used to place the *coronatum–athleta* zonal boundary. In the Grove Hill Borehole, the Acutistriatum Band is recognised between about 763.5 and 765.1 m; in the Brightling Borehole, the Comptoni Bed is recognised at about 685.2 m. A useful correlation between the two sequences is thereby established.

The Middle Oxford Clay consists of pale or medium grey and brownish grey slightly silty and silty calcareous mudstones. It is 21.0 m thick at Grove Hill (734.9 to 755.9 m) and 5.8 m thick at Brightling (676.7 to 682.5 m). The fauna, which is less abundant than that of the Lower Oxford Clay and which is commonly pyritised, includes the bivalves *Bositra buchii* (common at some levels), *Chlamys sp.*, *Discomiltha lirata*, *Grammatodon sp.*, *Isocyprina sp.*, *Meleagrinella braamburiensis* (common at some levels), nuculoids including *Mesosaccella morrisi* (common at some levels), *Nuculoma?* and *Palaeonucula sp.*, and *Thracia depressa*, together with the gastropods *Dicroloma* and *Procerithium*, the serpulid *Genicularia vertebralis*, and belemnite and wood fragments. The small coral *Trochocyathus magnevillianus*, known in abundance at this level elsewhere, for example at Calvert, Buckinghamshire (Callomon, 1968, p.285) and in the Warlingham Borehole, Surrey (Callomon and Cope, 1971, p.167), occurs at 745.2 m at Grove Hill and at 679.4 m at Brightling and provides a useful marker for correlation. The ammonite fauna, probably all belonging to the *athleta* Zone, is dominated by oppeliids and rarer perisphinctids, and includes, at Grove Hill, *Hecticoceras sp.* at 740.7 m, *Choffatia?* at 741.9 m, *Hec-*

ticoceras spp. at 743.7, 751.0, 751.3, 752.2, 752.9 and 754.4 m, *Binatisphinctes sp.* at 752.9 m, perisphinctid fragment at 753.5 m and *Kosmoceras sp.* at 753.8 m, and at Brightling, *Hecticoceras sp.* at 679.4, 679.7, 680.0, 680.6, 680.9 and 681.1 m.

The Upper Oxford Clay consists of predominantly pale grey, slightly silty, sparsely shelly, calcareous mudstone with pyritised trails and a mainly pyritised fauna. It is 36.9 m thick at Grove Hill (698.0 to 734.9 m) and 48.2 m thick at Brightling (628.5 to 676.7 m). The fauna includes the bivalves *Chlamys (Aequipecten)* cf. *scarburgense*, *Corbulomima?*, *Dacryomya?*, *Grammatodon sp.*, *Isognomon sp.*, *Modiolus sp.*, *Oxytoma sp.*, oyster fragments, *Palaeonucula sp.*, *Pinna sp.*, and *Thracia depressa*; crustacean fragments (common everywhere at this level in southern-central England), *Dicroloma*, belemnite and other shell fragments, debris and spat. Ammonites, mainly pyritised cardioceratids, indicative of the Lower Oxfordian *Quenstedtoceras mariae* Zone occur as follows: in Grove Hill, *Cardioceras sp.* at 701.7 m, about 702.9 m and 702.3 to 706.5 m, *Cardioceras (Scarburgiceras) praecordatum* at about 702.9 m, *C. (S.)* cf. *scarburgense* at 731.2 m, *Hecticoceras sp.* at 732.1 and 732.4 m, *Quenstedtoceras woodhamense* and *Q. (Pavloviceras?) sp.* at 732.4 m, and indeterminate cardioceratids at 702.3 to 706.5 m, 730.9, 731.5, 731.8 and 732.1 m; in Brightling, *Taramelliceras sp.* at 637.6 m, *?Quenstedtoceras mariae* at 639.2 m, *Cardioceras sp.* at 643.7 and 656.2 m, indeterminate cardioceratids at 661.1 and 662.3 m, *C. (Scarburgiceras) sp.* at 664.8 and *Hecticoceras sp.* at 665.1 m.

Corallian Beds

The Corallian Beds consist of silty and slightly silty, sparsely to moderately shelly, calcareous mudstones and siltstones, with argillaceous limestones containing ferruginous ooliths in the lower part of the sequence. They are 76.2 m thick in the Grove Hill Borehole (621.8 to 698.0 m) and 76.8 m thick in the Brightling Borehole (about 551.7 to 628.5 m).

In the mudstones, some of the fauna is partially pyritised, and pyritised trails are common. The fauna, based mainly on specimens from Grove Hill, includes the bivalves *Chlamys (Aequipecten) midas*, *Gresslya?*, *Isocyprina?*, *Isognomon* cf. *promytiloides*, *Modiolus?*, *Myophorella* cf. *clavellata*, *Nuculoma?*, oysters including *Deltoideum?*, *Gryphaea*, *Liostrea* and *Lopha*, *Pinna sp.*, *Placunopsis?*, *Plagiostoma?*, *Protocardia sp.*, *Pseudolimea sp.*, 'Tancredia'?, and *Thracia depressa*; with *Lingula sp.*, rhynchonellids and terebratulids (the last-named common at one level, 684.6 m, in Grove Hill); an echinoid, *?Nucleolites scutatus*, at 680.0 m in Grove Hill; *Dicroloma sp.* (common at one level, 648.0 to 650.1 m, in Grove Hill), crustacean fragments, and plant debris which is abundant on some surfaces. The ammonite fauna is dominated by cardioceratids, and the available specimens, although relatively few, allow the recognition of at least four of the standard Oxfordian ammonite zones based on cardioceratids (Sykes and Callomon, 1979). The ammonites are, in Grove Hill: *Ringsteadia sp.* at 625.8 to 631.5 m, (*Amoeboceras regulare* to *Amoeboceras rosenkrantzi* zones), *Amoeboceras* cf. *mansoni* at 640.1 m (*Amoeboceras serratum* Zone), *A.* cf. *regulare* at 643.1 m (*Amoeboceras serratum* Zone), *A.* cf. *glosense* at 650.4 and 651.1 m (*Amoeboceras glosense* Zone), *Perisphinctes sp.* at 651.1 m, *Amoeboceras sp.* at 652.9 m, *Cardioceras (Plasmatoceras?) sp.* at

681.8 and 683.1 m (*Cardioceras densiplicatum* Zone), *C. (Plasmatoceras) sp.* at 681.8 m (*densiplicatum* Zone) and *C. sp.* (ex gr. *densiplicatum-sowerbyi*) at 683.1 m (*densiplicatum* Zone); and in Brightling, *Cardioceras (Scarburgiceras) bukowskii* at 619.4 m (*Cardioceras cordatum* Zone, *Cardioceras bukowskii* Subzone), *C. (Scarburgiceras) sp.* at 620.0 m and *Cardioceras sp.* at 625.4 m (*cordatum* Zone). The lithofacies and ammonite fauna indicate that the sequence is much more like that of central and eastern England (Ampthill Clay – West Walton Beds) than the traditionally better known Corallian Beds of southern England.

Kimmeridge Clay

The thickness of the Kimmeridge Clay is estimated to be 286.5 m in the Grove Hill Borehole (335.3 to 621.8 m) and 407.2 m in the Brightling Borehole (144.5 to 551.7 m). The formation is conventionally divided into Lower and Upper, not least because of the substantial thickness of the formation in its type area (Cox and Gallois, 1981). The divisions are easily distinguishable on the basis of their ammonite faunas; the Lower Kimmeridge Clay is characterised by species of *Pictonia*, *Rasenia* and *Aulacostephanus*, and the Upper Kimmeridge Clay by species of *Pectinatites*, *Pavlovia* and related forms.

The Lower Kimmeridge Clay at Grove Hill and Brightling consists of siltstones, medium or dark grey mudstones, medium or pale grey sparsely shelly calcareous mudstones, and cementstones. The fauna includes bivalves, mainly fragmentary: *Camptonectes*, *Corbicellopsis?*, *Corbulomima?*, *Gervillella*, '*Lucina*'?, *Myophorella*, *Nuculoma?*, oysters including *Liostrea* and *Nanogyra* (particularly common and including *N. virgula* in the siltstones at about 530.4 to 531.9 m at Grove Hill), *Oxytoma*, *Placunopsis?*, *Protocardia?*, *Quenstedtia?*, *Thracia depressa*, *Trigonia*; small gastropods and *Lingula*. The scattered fragmentary ammonites are sufficient to give some stratigraphical control: in Grove Hill, *Sutneria sp.* and *Amoeboceras (Amoebites) sp.* at 534.3 to 538.6 m and a manuscript record of aptychi by Mr A. Templeman at 528.2 to 534.3 m (*Aulacostephanus eudoxus* Zone) and *Pictonia?* trans. to *Rasenia?* with *Amoeboceras (Amoebites)* at 592.2 to 595.0 m (*?Rasenia cymodoce* Zone); at Brightling, *Aulacostephanus* ex gr. *linealis* at 533.4 m (*Aulacostephanus mutabilis* Zone). The record of 'Crinoid stem fragments' in mudstone with limestone bands from about 502.9 to 518.2 m in Brightling suggests a correlation with the Pentacrinus Band of the *mutabilis* Zone (Gallois and Cox, 1976). In Grove Hill the base of the Kimmeridge Clay (and of the Kimmeridgian

Stage) is taken at about 621.8 m, at a downward change to silty and slightly silty calcareous mudstones with a partially pyritised fauna and pyritised trails. In Brightling, it is taken at about 551.7 m, where the log describes a downward change from smooth to silty, calcareous mudstones.

In the boreholes at Grove Hill and Brightling the Upper Kimmeridge Clay was either not recovered or not cored. However, the log of Brightling indicates that the sequence is predominantly of more or less calcareous and shelly mudstones with subsidiary limestone bands. The 'light brown limestone band' recorded at about 243.8 m has been correlated with the White Stone Band of Dorset (Downie *in* Worssam and Ivimey-Cook, 1971, pp.38 – 39); this stone band has been recognised in several boreholes in central and southern England, and marks the base of the Upper Kimmeridgian *Pectinatites pectinatus* Zone.

Portland Beds

The Portland Beds at Brightling and Grove Hill are apparently similar and comprise silty and/or calcareous mudstones, siltstones, argillaceous sandstones and argillaceous limestones. They are estimated to be 24.4 m thick at Grove Hill (310.9 to 335.3 m) and 31.7 m at Brightling (112.8 to 144.5 m).

At Grove Hill, the fauna, some of which is partially pyritised, consists of bivalves, commonly fragmentary, including *Anisocardia?*, *Camptonectes* cf. *lamellosus*, *C. morini*, *Entolium* cf. *orbiculare*, *Isognomon?*, *Modiolus autissiodorensis* and oysters (*Liostrea*, some epizoic, and *Nanogyra*), a rhynchonellid brachiopod, and incomplete or fragmentary ammonites amongst which are *?Galbanites okusensis* at 319.4 m, *Glaucolithites?* at 320.6 and 323.4 m, several *Glaucolithites sp.* at 324.3 to 328.9 m, *Epivirgatites* cf. *nikitini* at 329.5 m and *Epivirgatites?* at 324.3 to 329.2 m and 330.4 m. These ammonites allow the recognition of at least the two oldest Portlandian zones (Wimbledon and Cope, 1978) — from below, *Progalbanites albani* and *Glaucolithites glaucolithus* — and possibly also the next youngest *Galbanites okusensis* Zone. The boundary between the first two zones is taken at about 329.2 m. A possible position for the base of the *okusensis* Zone is at about 320.0 m, where there are small black angular to subangular phosphatic pebbles.

At Brightling Mine [677 219], large ammonites identified as *Glaucolithites glaucolithus* have been recorded from a 4-m thickness of Portland Beds below the lowest worked gypsum seam of the Purbeck Beds (Wimbledon and Hunt, 1983, p.268).

BMC

CHAPTER 3

Jurassic–Cretaceous: Purbeck Beds

GENERAL ACCOUNT

The Purbeck Beds are the oldest strata to crop out in the core of the Wealden anticlinorium. The results of studies of the three major inliers have been published (Howitt, 1964; Anderson and Bazley, 1971) and only a resumé of the conclusions of this work is given here (Figure 4). A BGS cored borehole at Broadoak [6195 2214], just north of the Lewes district, was sunk to establish the detailed stratigraphy (Lake and Holliday, 1978) and the sedimentary sequence proved is summarised as follows:

Thickness m

Greys Limestones Member: shaly mudstones with shelly partings and limestones containing *Neomiodon sp.*, subordinate calcareous siltstones (approximately equivalent to the Greys Limestones of previous authors) estimated at 26

Arenaceous Beds Member: coarsening-upward cycles ranging from shelly mudstone to sandstone 19

Thickness m

Cinder Bed Member: shelly mudstones with a limestone containing *Praeexogyra distorta* and *Myrene spp.* (correlated with the 'Cinder Bed' horizon of Dorset) 6

Plant and Bone Beds Member: shelly (*Neomiodon sp.*) mudstones and shelly limestones with rootlet horizons and greenish grey mudstones of seatearth aspect (approximately equivalent to the Plant and Bone Beds of Howitt, 1964) 14

Broadoak Calcareous Member: pyritic mudstones with abundant ostracods. Calcilutites (the Blues Limestones of previous authors) and pellet limestones abundant in upper beds. Shelly (*Neomiodon*) limestones in highest beds. Algal limestones. Limestones generally less abundant in lower beds. Calcilutites in basal beds. Some plant debris 56

Gypsiferous Beds Member: evaporite beds in four seams; the lowest and topmost are of gypsum, the intervening seams are mainly of anhydrite 16

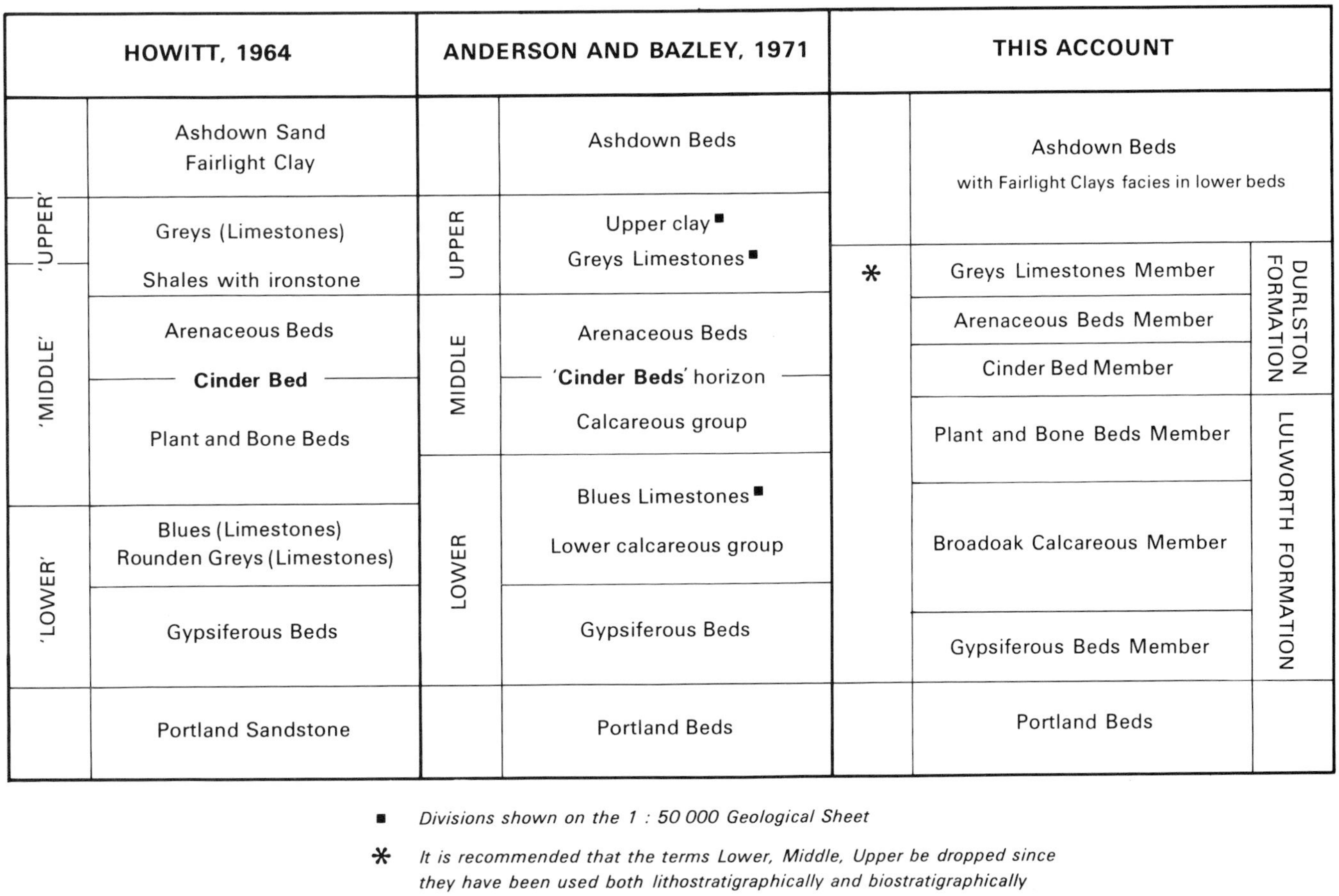

HOWITT, 1964		ANDERSON AND BAZLEY, 1971		THIS ACCOUNT		
	Ashdown Sand Fairlight Clay		Ashdown Beds		Ashdown Beds with Fairlight Clays facies in lower beds	
'UPPER'	Greys (Limestones) Shales with ironstone	UPPER	Upper clay ■ Greys Limestones ■	✱	Greys Limestones Member	DURLSTON FORMATION
'MIDDLE'	Arenaceous Beds **Cinder Bed** Plant and Bone Beds	MIDDLE	Arenaceous Beds 'Cinder Beds' horizon Calcareous group		Arenaceous Beds Member	
					Cinder Bed Member	
					Plant and Bone Beds Member	LULWORTH FORMATION
'LOWER'	Blues (Limestones) Rounden Greys (Limestones)	LOWER	Blues Limestones ■ Lower calcareous group		Broadoak Calcareous Member	
	Gypsiferous Beds		Gypsiferous Beds		Gypsiferous Beds Member	
	Portland Sandstone		Portland Beds		Portland Beds	

■   *Divisions shown on the 1 : 50 000 Geological Sheet*

✱   *It is recommended that the terms Lower, Middle, Upper be dropped since they have been used both lithostratigraphically and biostratigraphically*

Figure 4   Divisions of the Purbeck Beds

In the north-east of the Lewes district the Purbeck Beds are about 140 m thick, but they thin to at least 90 m at Pevensey in the south. The base of the formation (not exposed at the surface) is typically well-defined; beneath the lowest evaporite seam, thin laminated siltstones overlie bioturbated sandy Portland Beds. The divisions of the Purbeck Beds shown on the published 1:50 000 geological map are based mainly on the distribution of old pits for limestone and of occurrences of sandstones within the Arenaceous Beds. Figure 4 shows the approximate equivalence of the mapped divisions and the borehole sequence. Bazley (in Anderson and Bazley, 1971, p.11) distinguished a division of the Purbeck at the top of the sequence above the Greys Limestones, consisting of medium grey mudstones with subordinate silts, sandstones and nodular clay-ironstones (the 'upper clay' of Figure 4). This lithological unit is shown as Purbeck undifferentiated on the geological map, the top of the formation being taken, for field-mapping purposes, where arenaceous beds become dominant in the sequence. However, the evidence from boreholes indicates that this criterion is almost certainly subject to diachroneity and facies-changes: in places red-mottled clays of Fairlight Clays facies (p.16) or cyclic beds of mudstones and sandstones intervene. Therefore the top of the Purbeck is now taken at the top of the Greys Limestones Member, the overlying mudstones being regarded as part of the Ashdown Beds (Morter, 1984).

The macrofauna of the very fossiliferous upper part of the Purbeck Beds consists of the bivalves Neomiodon sp., Corbula sp. and Unio sp., with the gastropods Viviparus sp. Fish debris and plant debris are common. The faunal assemblages and inferred palaeosalinities have been reviewed by Morter (1984).

The chronostratigraphical divisions which have been applied to the Purbeck Beds do not necessarily coincide with major changes of lithology, which are probably diachronous. From studies of the ostracod faunas, Anderson (in Anderson and Bazley, 1971) divided the Purbeck Beds into Lower, Middle and Upper parts. There is no coincidence between these divisions and that of Casey (1963), who recognised a major marine influence in a limestone he correlated with the Cinder Bed of Dorset, which has generally been taken to mark the base of the Cretaceous System. However, recent studies of charophytes (fossil green algae) from boreholes in the Weald by Dr M. Feist of Montpelier University, France (personal communication), have revealed the presence of Globator maillardi near the base of the Broadoak Calcareous Member, including an occurrence at 111.50 m in the Broadoak Borehole. This species has been recorded from localities in the Jura, Portugal and Spain, and is considered to be characteristic of the Berriasian (lowest Cretaceous) charophyte zone of La Chaux (Grambast, 1974). It follows that the boundary between the Jurassic and the Cretaceous in the Weald may lie at the base of the Gypsiferous Beds Member (cf. Wimbledon and Hunt, 1983).

The evaporite deposits in the lowest 16 m of the Purbeck Beds are economically important (see p.92).

DETAILS

Dominantly calcareous mudstones exposed in the Dudwell valley between Tottingworth Farm and Rounden Wood have been extensively affected by landslips and valley bulges so that the measurement of continuous sections is not possible. Details of the important exposures have been published elsewhere (Anderson and Bazley, 1971, pp.13–17), and only a summary is presented here. Small exposures of beds referable to the Blues Limestones are present at the junctions of the tributaries from Milkhurst Wood [6260 2197], Ten Acre Wood [6308 2193] and Nine Acre Wood [6367 2184] with the main valley. Valley-bulged beds of the Greys Limestones Member are exposed in the bank of the River Dudwell near Poundsford Ford [6385 2167] (Figure 30). Beds within and above the Greys Limestones crop out in the tributary through Bingletts Wood [6227 2145]. The Blues Limestones have been extensively worked by bell-pitting in Rounden Wood [676 217]. Stream exposures in these beds occur between Rounden Wood and Ashen Wood [6762 2146]. Valley-bulge structures are described on p.89.

Exposures of historical interest are detailed by Topley (1875, pp.35–39).

The Grove Hill (Hellingly) Borehole [6008 1359] proved Purbeck Beds between 207 and 311 m depths, with abundant Praeexogyra distorta (Cinder Bed Member) from 230.1 to 231.0 m (Mr A. Templeman in MS). The upper boundary of the Purbeck was only approximately located on the basis of the abundance of shell debris below it. Calcareous ostracod shales occur above this boundary up to a level of about 189 m depth.

The sequence in the D'Arcy Oil Company's Pevensey No. 1 exploratory borehole [6265 0540], drilled at Hankham in 1938, may be considered in the light of the evidence from the Glynleigh Borehole [6085 0637] (p.103) and the following reclassification of the former record is now proposed:

	Thickness	Depth to base
	m	m
Faulted ground (Tunbridge Wells Sand and Wadhurst Clay) passing down to undisturbed Wadhurst Clay	64.6	64.6
Ashdown Beds	76.7	141.3
Purbeck Beds	88.4	229.7
Portland Beds	seen to 26.9	256.6

When compared with the Ashdown Beds sequence (211.6 m) proved at Fairlight [8592 1173] (Lake and Shephard-Thorn, in press) the thickness proved in the Pevensey No. 1 Borehole, which may be taken as a maximum figure allowing for the effects of faulting, shows considerable attenuation. Because the base of the Ashdown Beds is not easily interpreted from drillers' logs a further comparison may be drawn using the combined thickness of Ashdown Beds and Purbeck Beds. In the Pevensey Borehole this thickness is 165.1 m whereas at Fairlight it is 332.3 m. As the sedimentary facies at both localities are similar the observed variation is probably due to differential subsidence within the depositional basin.   RDL

CHAPTER 4

Cretaceous: Wealden

INTRODUCTION

The Wealden Series comprises sediments including mudstones, silts and siltstones, fine-grained sandstones either friable or lime-cemented, and thin conglomerates and pebble beds, together with thin shelly limestones and sideritic mudstones (clay-ironstones). In the Lewes district the series is divided into dominantly arenaceous or argillaceous formations as follows:

	Thickness m
WEALD CLAY	about 150
HASTINGS BEDS	
Tunbridge Wells Sand	
Upper Tunbridge Wells Sand	up to 100
Grinstead Clay	0 to ?10
Lower Tunbridge Wells Sand	25
Wadhurst Clay	30 to 55
Ashdown Beds	up to 210

The Grinstead Clay is present west of the longitude of Framfield, so that the Tunbridge Wells Sand can be subdivided in the north-west of the district. Two divisions of the Lower Tunbridge Wells Sand can also be recognised in the same area, the Ardingly Sandstone member above, comprising most of the massive sandstones in the sequence, and a more variable lower part.

The subdivisions of the Hastings Beds have been reviewed by Bristow and Bazley (1972), who defined their boundaries and listed the various reference sections. The subdivisions used here depend on lithological characters recognisable in the field, and are probably diachronous.

The above stratigraphy is based on the pioneer studies of Mantell (1822), Drew (1861), Topley (1875) and his colleagues. In more recent years, Allen (1949–1981) has made a major contribution to the understanding of the sedimentology and stratigraphy of these rocks; in particular, he has demonstrated the presence of important marker beds within the sequence. Two of these, the Top Ashdown Pebble Bed and the Top Lower Tunbridge Wells Pebble Bed, define formation boundaries. These pebble beds, which characteristically overlie fairly thick sandstones, mark transgressive phases within the sequence (Allen, 1949). The red and green mottled clays which occur within or near the tops of the clay formations may indicate shallowing of the basin waters (to permit oxidising conditions) prior to the influx of coarser sediment.

Allen (1959) recognised three major cyclothems in the Wealden Series, essentially pairs of arenaceous and argillaceous formations. These cyclothems comprised the Ashdown Beds and Wadhurst Clay, the Lower Tunbridge Wells Sand and Grinstead Clay, and the Upper Tunbridge Wells Sand and Weald Clay. Minor cyclothems were identified within this sequence, notably where the Grinstead Clay was divided by the Cuckfield Stone to the north-west of the district. An idealised cyclothem may be summarised as follows:

— Transition or sharp erosive break —
Clays and thin limestones, commonly with *Equisetites* soil beds, *Neomiodon sp.* shell-beds and ostracods; thin lenticular sandstones, particularly near the base; dark ostracod-rich clays tend to dominate towards the top | Argillaceous part of cyclothem
Pebble-beds, thin, graded |
— Sharp break with erosion —
Sandstones dominant, massive near the top, locally replaced or overlain by argillaceous sandy siltstone | Arenaceous part of cyclothem
Siltstones and silty clays with subordinate sandstones, tending to be more argillaceous downwards |

Allen (1959) thought the phases of each cyclothem represented the growth and subsidence respectively of successive delta complexes, in a predominantly freshwater environment. However, he subsequently concluded that they were sympathetic responses both to oscillations of Neocomian sea level (Allen, 1959, pp.341–342) and to periodic tectonic uplift of the source areas (Allen, 1975). In this latter paper Allen proposed a model in which alluvial and lagoonal mudplains were at times invaded by braided rivers; these deposited sheets of sandy sediment which may have grown into coalescent gravelly alluvial fans. In 1981, Allen suggested that a distal meander plain zone might also have been present within the basin. Subsequently, Stewart (1983), from a study of the Ashdown Beds of the Hastings coast-section, observed that meandering streams with high suspension-loads were responsible for the laterally accreting channel structures present.

Allen has made detailed studies of the origin of the sediments (1949; 1954; 1959; 1962) and suggested the London Platform–Ardennes land mass (the Brabant Massif) as a primary source of much of the material. Somewhat paradoxically, studies of faunal provinces and sedimentary facies variations indicate that marine influences which affected the Wealden basin came from the Boreal sea, also to the north (Kaye, 1966; Lake and Thurrell, 1974; Allen, 1981). RDL

Ostracods are the most abundant fossils in the Wealden and the ostracod genus *Cypridea*, probably living by preference under mixohaline conditions, includes a large number of species and subspecies and so is ideally suited for use in the division of the strata (Anderson, 1940; 1985). The sequence of ostracod faunas in the Purbeck and Wealden strata is characterised by repeated alternations between assemblages mainly composed of species of *Cypridea* and those in which the dominant forms are species of genera other than *Cypridea*. In the latter case many genera are represented but few by more than two or three species.

The assumption is made that all the species of *Cypridea* lived in a similar but not necessarily identical ecological niche and that the salinity of the water in which they lived was one of the major factors determining their distribution. Because the *Cypridea* and non-*Cypridea* assemblages appear to be antipathetic it is further assumed that the non-*Cypridea* assemblages preferred ecological conditions different from those favoured by the species of *Cypridea* and also that this difference was primarily one of salinity. These variations in the composition of the fauna are thought to have been controlled by alternating periods of lighter and heavier rainfall, which brought about changes in the salinity of the water in which the ostracods lived. These alternations, referred to as 'faunicycles', form a basis for correlation throughout southern England. In general the S-phase (marine or quasi-marine) faunas, usually consisting of long-range species, cannot be individually distinguished. But the rapidly changing C-phase (less saline) faunas which separate them are generally distinctive and make recognition of the cycles possible. Secondly, the C-phase faunas form a succession of assemblages the composition of which is clearly controlled by ecological factors different from those which determined the faunicycles. Only rarely do changes in the composition of the assemblages coincide with the S-phases of a faunicycle, and thus they form a useful second approach to correlation. Unlike the faunicycles, the assemblages appear to be influenced in some degree by local conditions.

In the Wadhurst Clay the S-phase forms include species of *Darwinula*, *Mantelliana*, *Orthonotacythere*, *Rhinocypris*, *Theriosynoecum* and *Timiriasevia*. Eleven faunicycles were recognised in the Wadhurst Clay of the Glynleigh Borehole and 29 in the Weald Clay of the Ripe Borehole. In the Weald Clay the marine S-phase faunas mainly comprise species of *Theriosynoecum* and *Miocytheridea*. The species of *Cypridea* in the C-phase form a rapidly changing complex, so that many of the faunicycles can be identified by the Cypridean assemblage alone.

That a particular ostracod species has been chosen to be the name fossil of a zone does not imply that it is present only in the strata comprising that zone, but rather that it is abundant and is an important member of the ostracod assemblages found there.

The zones listed below are all recognisable by characteristic assemblages of ostracods named from species of the genus *Cypridea*.

Zone	Lithological divisions
C. valdensis	
C. clavata	
C. marina	Weald Clay
C. tuberculata	
C. dorsispinata	
C. aculeata	Hastings Beds (Tunbridge Wells Sand including Grinstead Clay, upper and middle Wadhurst Clay)
C. paulsgrovensis	Hastings Beds (lower Wadhurst Clay)
C. brevirostrata	Hastings Beds (Ashdown Beds)

In the *C. paulsgrovensis* Zone few species are represented, though individuals may be numerous; apart from the name fossil, only *C. laevigata* and *C. tuberculata* are common. The *C. aculeata* Zone contains an ostracod fauna richer and more varied than in any other part of the Wealden. The name fossil itself is a variable species and together with *C. bispinosa* dominates the fauna. Two other distinctive species, *C. recta* and *C. melvillei*, are characteristic of the zone.

Detailed information on the Wealden faunas has been provided in Anderson (1962; 1985) and Anderson and others (1967). RDL,FWA

STRATIGRAPHY

Ashdown Beds

The outcrop of the Ashdown Beds occupies an extensive area in the north-eastern quadrant of the district. The Ashdown Beds consist mainly of silts with subordinate sands and clays. The clay content is greater in the lower part of the formation, and here the clays are locally sufficiently thick to form mappable units (the Fairlight Clays facies). In the upper part of the formation, clay beds are generally impersistent. Because of the high proportion of silts in the Ashdown Beds, many logs made by well-sinkers show 'clay' through much of the formation, since the silts have a clayey consistency when wet. However, cored boreholes in the district have penetrated the upper beds of the formation and have provided reliable and detailed information on the lithologies (Lake and Young, 1978; see also Lake and Thurrell, 1974). Grading between the sands, silts and clays is observed; coarsening-upwards cycles occur locally. Plant debris and rootlets are common, associated with sphaerosiderite.

The Top Ashdown Sandstone (Plate 2) is a persistent fine- to medium-grained sandstone at the top of the formation and ranges from 1.2 to 8 m in thickness. The Top Ashdown Pebble Bed above it is generally thin and impersistent in this district, and a gradational sequence of interbedded sands and clays is commonly present at about this level, in contrast to the relatively sharp lithological change observed elsewhere (p.17). The base of the pebble bed marks the base of the Wadhurst Clay, and has been taken to represent the transgression of this formation across the Ashdown Beds (Allen, 1959). In genetic terms, however, it is possibly preferable to regard the Top Ashdown Sandstone of this district as a transgressive lag-deposit, and the pebble bed as merely its last component.

The thickness of the Ashdown Beds ranges from about 210 m in the north-east of the district to about 80 m at Pevensey. The Grove Hill Borehole [6008 1359] proved 174 m of strata that have been assigned to the Ashdown Beds, but this figure is approximate owing to the difficulties of defining the base of the formation (p.14).

Fossils are generally rare within the Ashdown Beds, although in boreholes estheriids have been observed, typically below some of the sandstones, more particularly beneath the Top Ashdown Sandstone.

Wadhurst Clay

The greater part of the Wadhurst Clay consists of medium grey mudstones which weather to heavy, ochreous and greenish grey clays at the surface. Subordinate sandstones, siltstones, siderite-mudstones and shelly limestones also occur. Locally the thin arenaceous beds are cemented by calcite to form hard rock bands termed 'Tilgate Stone' (see, for example, White, 1926, pp.24–25). The formation ranges in thickness from 55 m in the Uckfield area to 30 m in the south-east (Glynleigh Borehole); 50 m were proved in the East Hoathly Borehole (Lake and Young, 1978). In the Bexhill area to the east, thicknesses of the order of 39 m have been recorded, whereas thicknesses of over 70 m have been proved in boreholes in the Tunbridge Wells district to the north (Bristow and Bazley, 1972).

Plate 2   Top Ashdown Sandstone at Waldron (A 12293): festoon-bedded, fine-grained sandstone

Lake and Young (1978) have shown from boreholes in the Lewes and Hastings districts (Figure 5) that here the formation falls into broad lithological divisions as follows:

5   Red and green mottled mudstones
4   Grey shelly mudstones, commonly with a conglomeratic horizon at base
3   Cyclic sequence — see below
2   Grey shelly mudstones
1   Passage series from Ashdown Beds

Above the Top Ashdown Pebble Bed, interbedded silt-stones and clays pass up into a soil bed with roots and rhizomes of *Equisetites lyelli* in position of growth. Shell-beds with *Neomiodon medius* ('*Cyrena*') succeed thin dark clays above the soil bed. Recent cored boreholes have shown that soil beds or their equivalent rootlet horizons occur through-out the formation, although the basal Brede *E. lyelli* Soil Bed of Allen (1941; 1947) is generally the best preserved.

This sequence (divisions 1 and 2) is comparable to that outlined by Allen (see p.15). The Top Ashdown Pebble Bed is locally absent, particularly in the eastern part of the district (i.e. east of Horam) where the junction between the Ashdown Beds and the Wadhurst Clay is transitional, with a passage group of interbedded sands and clays. Division 2 is

generally less silty than the other divisions and contains abundant partings of bivalves and ostracods; gastropod re-mains are less common. Rootlet horizons are few in division 2, although greenish grey 'seatearth' lithologies occur local-ly. The cyclic sequence (3) typically comprises cycles show-ing the following lithological units and associations:

Mudstone, greenish grey, calcareous (seatearth), passing down to
Mudstone, medium grey, with *Neomiodon sp*. and ostracods, passing down to
Mudstone, medium grey, with silt laminae and rootlets
Sharp base
Mudstone, greenish grey, of preceding cycle

The greenish grey mudstones are commonly thinner than the underlying units. Mud pellet layers, indicative of pene-contemporaneous sediment reworking, are present in divi-sions 2 and 3.

Division 4 consists predominantly of medium grey mud-stones with *Neomiodon sp*. The conglomeratic horizon is in places a bed containing ironstone and mudstone clasts, and elsewhere a calcareous sandstone with fragments of teeth and bone, and ironstone and mudstone pellets.

The red and green mottled mudstones (5), which weather to an overall red colour at the surface, are ubiquitous near the top of the Wadhurst Clay and form a useful marker for

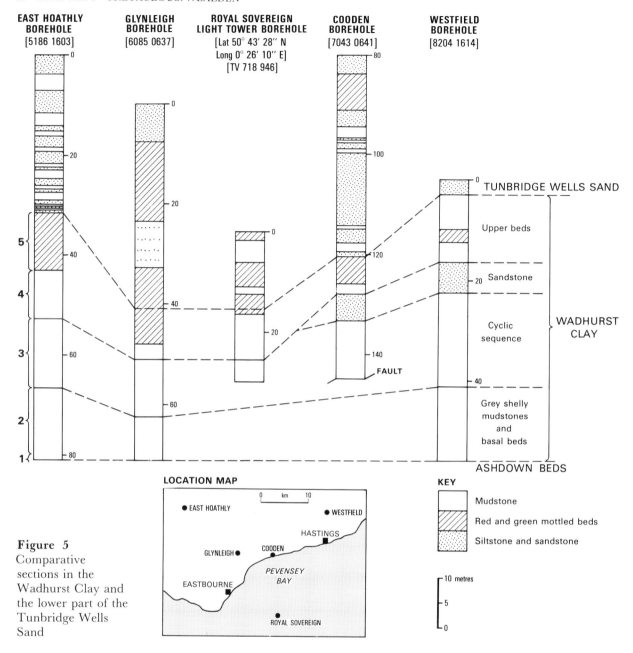

Figure 5
Comparative sections in the Wadhurst Clay and the lower part of the Tunbridge Wells Sand

mapping purposes. The five divisions, with the exception of the basal silty beds and the red-weathering mudstones, cannot be distinguished at outcrop because most of the mudstones weather to a uniform greenish grey colour. Locally, however, grey clays can be proved by augering.

Ironstone (siderite-mudstone) occurs as nodular or tabular beds throughout the formation and formerly provided the main source of iron ore in the district. The distribution of old pits dug for the Wealden iron industry indicates that ironstone beds are most abundant in the basal part of the formation and hence were the most important economically (see p. 91).

No horizon equivalent to the Telham bone-bed of the lower Wadhurst Clay, which occurs in the Hastings district to the east (Topley, 1875; White, 1926; Allen, 1949), was recognised during the present survey.

Tunbridge Wells Sand

The Tunbridge Wells Sand is a complex cyclic sequence of siltstones with subordinate sandstones and clays. The cycles, best developed in the Lower Tunbridge Wells Sand and equivalent beds throughout the district, mainly fine upwards. Rootlets and plant debris are abundant in the whole succession, lignite being particularly common above sharp lithological contacts and suggesting that the depositional environment was dominantly that of alluvial channels with extensive vegetation. Local pellet-beds represent lag-gravels. The base of the Tunbridge Wells Sand is generally sharp, and a well-defined spring-line is commonly present at the junction with the Wadhurst Clay, but the top is gradational.

Two differing successions are present in the Lewes district (p. 15). West of Framfield the Lower Tunbridge Wells

Plate 3   Ardingly Sandstone at Founthill (A 2945)

Sand, with the Ardingly Sandstone at its top, is separated from the Upper Tunbridge Wells Sand by the Grinstead Clay. In the eastern part of the district, however, no median clay has been mapped.

The Ardingly Sandstone (Plate 3), below the Grinstead Clay, is a fine- to medium-grained massive sandstone up to 12 m thick; it shows a tendency to coarsen upwards and contains slump structures and pellet-beds. The Top Lower Tunbridge Wells Pebble Bed is locally present and this rests disconformably on the Ardingly Sandstone. This sequence of beds is similar to that at the junction of the Ashdown Beds and Wadhurst Clay (p.16).

The Grinstead Clay consists mainly of dark grey shaly clays, with subordinate thin lenses of sandstone and siltstone in the basal beds. *Unio*, *Neomiodon*, *Viviparus* and ostracods are present. The basal beds are penetrated by *Equisetites lyelli* rootlets, constituting a soil bed (Allen, 1959). Red mottling is general near the top of the formation.

The Upper Tunbridge Wells Sand and equivalent beds consist of a lower group of red and grey mottled silts and clays with pellet-beds and local thick channel-floor sandstones; the upper part mainly comprises alternations of silts and silty clays with subordinate thin sandstones.

In the eastern part of the district thick sandstones are generally confined to the middle part of the Tunbridge Wells Sand (Figure 6). Cyclic beds are present beneath, and alternations of silts and silty clays with subordinate thin sandstones occur above.

Westwards of Herstmonceux a sandstone about 18 m above the base of the Tunbridge Wells Sand (Figure 7) may be the correlative of the Ardingly Sandstone. The sandstone appears to have been laterally persistent, although it is locally absent because of channelling. Above it lie red-mottled clayey beds which are laterally more extensive and occupy a broad 'zone' in the middle of the sub-group. In places, however, the mottled silts and clays pass into grey laminated beds. The former, proved in the Cooden Borehole [7043 0641] just east of the Lewes district, have been correlated with similar beds above the Grinstead Clay in the west (Lake, 1975a). If this correlation is correct it would seem that the Ardingly Sandstone and Grinstead Clay, which are apparently genetically linked, were both originally deposited in the eastern part of this district; a subsequent diachronous phase of emergence of the Wealden Basin, associated with renewed channelling, caused the oxidation and/or erosion of the Grinstead Clay here.

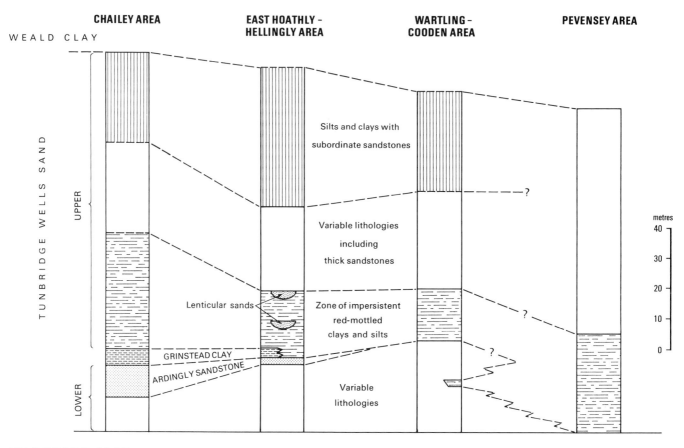

Figure 6 Schematic diagram to show the relationships of the subdivisions of the Tunbridge Wells Sand

In the Pevensey area, the Glynleigh Borehole [6085 0637] showed that red mottling in dominantly clay lithologies extends down to the top of the Wadhurst Clay. This phenomenon appears to be restricted to the southern area of the district. A comparable succession was observed in the Royal Sovereign Light Tower Borehole (Latitude 50°43′28″N, Longitude 0°26′10″E) offshore to the southeast (Figure 5).

The boundary between the Tunbridge Wells Sand and the Weald Clay (Figure 8) is generally gradational, so that precise definition of the top of the former and determination of its thickness are rarely possible. The main sandstone group of the Tunbridge Wells Sand (Figures 6 and 8) is generally the first distinctive unit recognised in wells penetrating this sequence and the thickness of the Weald Clay in boreholes is often over-estimated at the expense of the beds below. The Cooden Borehole proved the full thickness of the Tunbridge Wells Sand as 107 m. An overall thickening of the sub-group westward and northward is indicated by the Cuckfield No. 1 Borehole in the Horsham district, which proved 154 m of Tunbridge Wells Sand. A well [3832 1720] near Chailey, just west of the Lewes district, has a record which is difficult to classify but suggests approximate thicknesses of 65 m of Weald Clay overlying 125 m of Tunbridge Wells Sand.

Weald Clay

The Weald Clay, which crops out over a broad belt of undulating lowland extending from the western edge of the Lewes district to the coast near Pevensey, is predominantly argillaceous with very subordinate sand, silt, limestone and ironstone. The unweathered sediments are pale to dark grey, brown, greenish grey and red, but within 10 m of the surface shades of yellow, fawn and brown are assumed because of the alteration of iron compounds; clays that are mottled with red and green in the unweathered state tend to become an overall deep red, and 'catsbrain' mottling (weak red mottling on pale greenish grey clays) is developed by some green clays.

Since the base of the Weald Clay is generally poorly defined, certain well logs may include clayey transition beds of the Tunbridge Wells Sand in the Weald Clay and hence exaggerate the thickness of the latter at the expense of the former. Nevertheless, borehole evidence indicates a thickness of around 150 m for the formation.

The fauna of the Weald Clay includes ostracods, species of the gastropod *Viviparus* and the bivalve *Filosina* ('*Cyrena*'). Fragments of reptile bones, fish and plants also occur, commonly associated with the arenaceous beds. Ostracods are generally most abundant and form the basis of a zonal

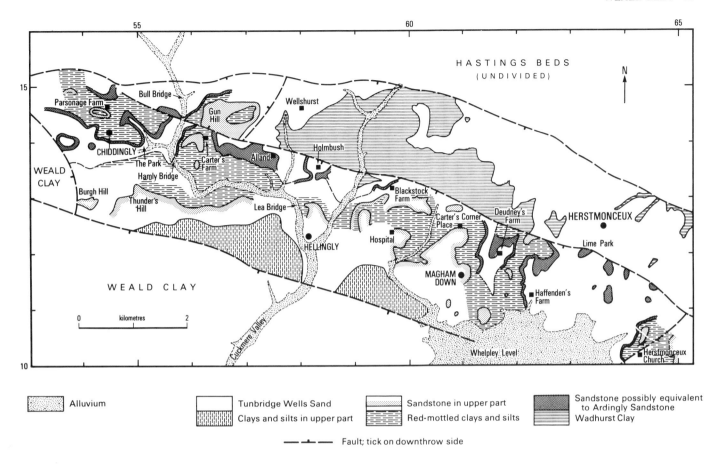

55 60 65

15

HASTINGS BEDS
(UNDIVIDED)

N

Parsonage Farm
Bull Bridge
Gun Hill
Wellshurst
CHIDDINGLY
The Park
Carter's Farm
Alland
Holmbush
WEALD CLAY
Hamly Bridge
Burgh Hill
Thunder's Hill
Lea Bridge
Blackstock Farm
Carter's Corner Place
Deudney's Farm
HERSTMONCEUX
HELLINGLY
Hospital
Lime Park
WEALD CLAY
MAGHAM DOWN
Haffenden's Farm
0 kilometres 2
Cuckmere Valley
Whelpley Level
Herstmonceux Church

10

| Alluvium | | Tunbridge Wells Sand | | Sandstone in upper part | | Sandstone possibly equivalent to Ardingly Sandstone |
| | | Clays and silts in upper part | | Red-mottled clays and silts | | Wadhurst Clay |

— · — · — Fault; tick on downthrow side

Figure 7 Sketch-map showing the lithologies of the Tunbridge Wells Sand in the Hellingly area

classification (p.16); most of the ostracods are freshwater forms, quasi-marine types being restricted to certain levels.

In the Lewes district shell-beds are rarely sufficiently well developed to form limestones, so that limestone soil-brash is exceptional. Large-'*Paludina*' limestone, also locally termed Petworth Marble, Sussex Marble or Laughton Stone, is composed of closely aggregated shells of the globose gastropod *Viviparus fluviorum* (J. Sowerby *non* de Montfort) and locally is thick enough to have been worked as a building-stone (Plate 4). In places the Large-'*Paludina*' limestone gives rise to moderately defined topographical features, although scarp and dip-slope features comparable with those in the Haslemere district for example (Thurrell and others, 1968, p.21) are rare. The arenaceous beds in the Weald Clay generally occur as sheet-like seams of friable fine-grained sand, locally resting on irregular channelled surfaces. Carbonate-cemented sandstones and siltstones are less common.

Seams of clay weathered to a homogeneous red colour are commonly associated with the sand bodies (Topley, 1875, p.97) and these occur at well-defined horizons, reflecting emergent, oxidising conditions in the depositional basin. In boreholes these clays are mottled with red, purple, brown and green and show vertical filamentous mottling suggesting the former presence of rootlets. Thurrell and others (1968, pp.22–23) proposed an idealised sedimentary cycle for these beds:

Sand beds
Sharp base
Clays, greenish grey, with limestones
Clays, grey
Clays, silty, red and grey-mottled, alternating with
Sand beds, on sharp base

In the Lewes district such cycles are apparently poorly developed and the sequence of associated beds in descending order appears to be:

Sand bed
Clay, silty, grey, locally red-mottled
Clay, red
Clay, greyish green or khaki, calcareous
Clays, grey
Clay, silty, red and grey-mottled
Sand bed

The red clays are useful markers for field mapping purposes; where they are absent the associated greyish green clays have been traced. Both of these lithologies comprise the marker clays shown on the 1:50 000 geological sheet.

SUBDIVISIONS OF THE WEALD CLAY

In the Haslemere district, Thurrell and others (1968, p.24) used a scheme for distinguishing lithological marker horizons modified after Topley (1875, p.102), namely:

5 et seq. Alternating sands and Large-'*Paludina*' limestones
4 Large-'*Paludina*' limestones

3 Sandstone
2 Small-*'Paludina'* and *'Cyrena'* limestones
1 Horsham Stone

In the Lewes district Bed 1 and beds above Bed 4 are absent and the sequence to be described (Figure 8) will follow the order:

4 Large-*'Paludina'* limestone
3 Sands and red clays
2 Small-*'Paludina'* and *'Cyrena'* limestones

Reeves (1958) proposed a threefold division of the Weald Clay, in which the oldest and youngest red clays separated groups of strata characterised by their dominant colours at outcrop, viz: Group I (lowest)—buff-grey; Group II—red; Group III—yellow. Since the red clays are impersistent and are used in conjunction with the greyish green clays for field-mapping purposes, these two schemes are not equivalent, although the oldest red clay occurs at about the top of the sequence of *'Cyrena'* limestones and the youngest occurs in association with the Large-*'Paludina'* limestone.

Owing to the impersistent nature of the Large-*'Paludina'* limestones and the strike-faulted outcrop of the Weald Clay it is not always possible to distinguish between the various clay marker units, particularly since few lower limestones (of Bed 2) have been recorded at outcrop. Boreholes at Ripe, Hailsham and Hampden Park (in the Eastbourne district) have, however, provided cored sections of the Weald Clay (Figure 8). Field-mapping has demonstrated that the Large-*'Paludina'* limestone and/or its marker clay equivalent is continuous throughout the area. The individual clay markers comprising Bed 3 appear to be impersistent but their overall presence as a group has been confirmed. The associated sands of Bed 3 are usually either lenticular or too thin to be mapped over any great distance.

Detailed descriptions of the cored boreholes in the Weald Clay have been published elsewhere (Lake and Young, 1978). The summary log of the Ripe Borehole [5059 1052], which displays the full sequence, records the following units:

	Thickness m
Mudstones characterised by:	
1 alternations of silt-striped and ostracod-bearing grey beds; some bivalve partings; brackish-water fauna in the upper part	16.23
2 alternations of silt-striped and grey beds with fish debris; ostracods less common	20.76
3 cyclic increase of silt lamination downward within units on eroded bases; grey and greyish green	11.72
4 Large-*'Paludina'* in a marly lithology, red and green-mottled at the top; *Filosina* and estheriids at base	5.50
5 fish and ostracod partings with calcareous siltstones in grey beds	1.23
6 dominantly silt-striped, grey beds with rootlets; some olive-grey and brown colouration, some cyclic development as in unit 3	16.71
7 red, green and purple mottling with sandy beds	8.56
8 silt-striped grey beds with rootlets; some colour mottles	5.67
9 green colour with sandy beds	3.33
10 silt-striped grey beds with plant debris	5.62
11 red, green and purple mottling with sand intercalations alternating with silt-striped beds in the lower parts; estheriids	9.80

	Thickness m
12 alternations of silt-striped and ostracod-bearing beds; locally green	8.18
13 red, green and brown mottling	1.55
14 alternations of greenish grey beds and silt-striped beds; rootlets	2.77
15 Small-*'Paludina'* and *'Filosina'*; local intercalations of silty beds	23.55
16 silty, laminated or bioturbated grey beds; sideritic beds; locally brown; almost barren of fauna	17.68
Weald Clay thickness	158.86

In addition, field mapping has demonstrated the presence of sand lenticles above unit 4.

In the above sequence, unit 4 is Bed 4 of the Lewes district sequence, units 7 to 13 constitute Bed 3 and unit 15 is equivalent to Bed 2. In the Hampden Park Borehole (Lake and Young, 1978), unit 1 was absent, probably as the result of a structural unconformity at the base of the Lower Greensand (see Figure 8). The following fossils have been identified in unit 1: cf. *Cuneocorbula arkelli*, *Filosina gregaria*, *F.* cf. *membranacea*, *Nemocardium (Pratulum) ibbetsoni* and *Praeexogyra* cf. *distorta*.

Dr M. Feist has identified the following charophytes in samples of Weald Clay from BGS boreholes in this district. In the Hailsham Borehole: *Triclypella calcitrapa* at depths of 16.50 to 17.00 m and 38.00 to 38.50 m; *Clypeator combei* at 17.00 to 17.50 m. In the Ripe Borehole: *T. calcitrapa* at 154.50 to 155.00 m and *C. combei* at 159.50 to 160.00 m. She reports that these species are Lower Barremian forms and that their occurrences accord with the correlation discussed below (Figure 8).

In the Lewes district, all but the lowest beds of the Weald Clay crop out consistently along the strike. The basal sequence is for the most part faulted out against the Tunbridge Wells Sand along the northern boundary, but in the vicinity of the Pevensey Levels it is present although affected by faulting. The higher beds are commonly repeated by strike-faults and associated flexures.

Twelve samples of representative lithologies of the Weald Clay in the Ripe Borehole have been analysed for their particle size and clay mineral content by Messrs G. E. Strong and R. J. Merriman respectively. They have found that the clay fraction (defined as 2 microns or less) ranges from 10 per cent in a brown and green-mottled seatearth to 52 per cent in a typical grey silty mudstone. In general, however, the siltier beds contain between 20 and 40 per cent of clay. The colour-mottled beds were found to contain significant proportions of sand (up to 16 per cent). X-ray diffraction analysis revealed that the clay minerals present are typically: clay-mica, at least 30 per cent; kaolinite, 10 to 20 per cent; chlorite, up to 15 per cent. Up to 10 per cent of montmorillonite was present in two samples from the highest part of the Weald Clay (unit 1).

REGIONAL CORRELATION

The subdivision of the Weald Clay in the Hailsham area does not fall readily into the sequence of Topley (1875) as modified by Thurrell and others (1968). Although Bed 4, Large-*'Paludina'* limestone or the associated red-mottled clay, occurs throughout the Lewes district, the definition of

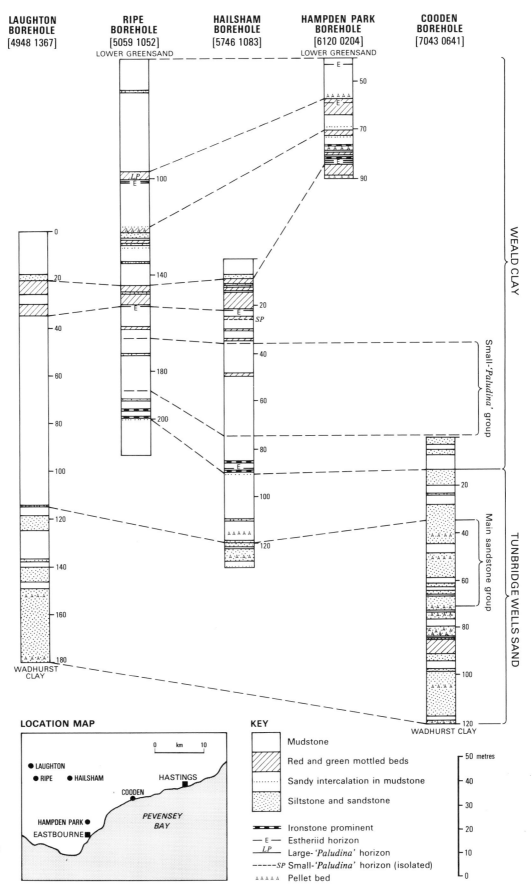

Figure 8 The Tunbridge Wells Sand–Weald Clay sequence

Bed 2 is imprecise, depending on the patchy presence of what may be regarded as a facies-fauna. The occurrence of neither Small-*'Paludina'*, which is used to define Bed 2, nor *Filosina* is strictly comparable in the Ripe and Hailsham boreholes. Correlations involving Bed 2 are therefore applied only in a broad 'zonal' sense (Figure 8).

Studies of the ostracod faunas show that Bed 4 is equivalent to the Gilmans faunicycle characterised by *Cypridea bogdenensis* (see Worssam and Ivimey-Cook, 1971, p.126) and indicate a lack of precision of correlations within Bed 2. The occurrence of the estheriid *Cyzicus (Lioestheria) subquadrata* above Bed 2 has been used for correlation purposes (Figure 8). This record appears to be directly linked to the Bonnington cycle in the *Cypridea clavata* beds (Dr F. W. Anderson, personal communication).

Correlations using the *bogdenensis* fauna indicate that Bed 4 of the Lewes district is equivalent to the Bed 4 which is present at Gilman's Brickpit [0855 2471], Billingshurst (Anderson *in* Thurrell and others, 1968, p.29), and Bed 6 of the Warlingham Borehole (Anderson *in* Worssam and Ivimey-Cook, 1971, p.126). At the latter site this was the only Large-*'Paludina'* limestone present in the sequence. This correlation is considered to be the most valid at the time of writing, although Dr F. W. Anderson (personal communication) extended the range of *C. bogdenensis* to include the Gilmans and succeeding Ditchling faunicycles (see Anderson *in* Worssam and Ivimey-Cook, 1971, p.126). This species was found in the middle of the *C. clavata* Zone at Warlingham and Ripe and near the base of the zone at Billingshurst.

RDL

DETAILS

Ashdown Beds

For convenience, some localities which expose the Top Ashdown Pebble Bed are included here.

In the area around Framfield [495 205] the Ashdown Beds outcrop is fault-bounded except in the vicinity of the village, where the formation is conformably overlain by the Wadhurst Clay. Exposures are poor in the main part of the formation, being limited to overgrown sections in sunken lanes. Augering revealed a sequence of fine-grained sands and silts. The Top Ashdown Pebble Bed is exposed periodically in the village cemetery; it has been noted by Professor P. Allen (personal communication) in an old quarry [4993 2064] nearby.

CRB

On the west side of the A265 road, yellow fine-grained sands assumed to belong to the Top Ashdown Sandstone have been excavated from degraded pits [5183 1959] 300 m south of Pembroke Manor. Approximately 225 m east of this, in the grounds of Cider House Farm [5216 1956], 1.5 m of massive sandstone were exposed. In an old pit [5264 1968] 700 m SW of Dower House Farm, 4.6 m of Top Ashdown Sandstone were exposed, the top 1.8 m being pale fawn cross-bedded fine-grained sandstone.

In the bed of the small stream [5321 1791] approximately 450 m NE of Hope's Farm the Top Ashdown Sandstone and basal Wadhurst Clay are exposed over a distance of 20 to 30 m; the section may be summarised as follows:

	Thickness m
WADHURST CLAY	
Silty mudstone and siltstone, interlaminated, with some shelly ironstones	0.9
Sandstone, coarse-grained, with quartz grains up to 2 mm in diameter (Top Ashdown Pebble Bed)	0.08
ASHDOWN BEDS	
Siltstones and clays, grey, cross-bedded	0.3
Sandstone, pale grey, massive	1.0

A small degraded pit [5327 1857] south of the lane at the southern end of Hawkhurst Common Wood showed up to 1.8 m of pale fawn, cross-bedded, poorly cemented fine-grained Top Ashdown Sandstone. A poorly exposed section in the north side of the lane [5323 1857] appeared to show the Wadhurst Clay resting on the Ashdown Beds with no intervening pebble bed. Here 0.6 m of massive, pale grey sandstone was overlain by 1.2 m of fawnish grey silty clays. Between Hawkhurst Common and Waldron [550 192] augering showed the general persistence of the Top Ashdown Pebble Bed.

A track-cutting [5405 1863] about 100 m south of Heronsdale Manor showed:

	Thickness m
WADHURST CLAY	
Silty clays	0.3
Sandstone, coarse-grained, yellow, with quartz and chert grains up to 3 mm; ripple-marked top (Top Ashdown Pebble Bed)	0.08
ASHDOWN BEDS	
Silts and medium- to coarse-grained sands, interbedded, pale grey, with scattered 2-mm sand grains	0.5
Sandstone, fine-grained, pale grey, relatively massive	0.3

The Top Ashdown Sandstone is exposed in long sections in the sunken lanes in and around Waldron village (Plate 2). In the lane leading south from the village up to 1.8 m of pale fawn silty sandstone commonly showing small-scale festoon-bedding is exposed for 150 m from about 130 m south of the War Memorial. Immediately south of this an old pit [5496 1896], in which the house known as 'The Rocks' stands, exposes 2.4 m of the same sandstone. A further sunken lane section extends for 120 m SE from the lane junction near Tullaghmore [5506 1878], where up to 3.7 m of Top Ashdown Sandstone are exposed.

White (1926, p.17) recorded lignite from Brown's Lane, Waldron, and from Waldron Gill, 2.4 km east of Waldron village. A clay bed within the Ashdown Beds was traced by augering for about 750 m along the slopes of Whitehouse Gill, near the lane junction [5558 1951] about 650 m ENE of Waldron. On the south side of the lane a newly cleared ditch extending about 140 m ENE of this lane junction provided the following section:

	Thickness m
Silty sands, passing down to silty clays	0.9
Clay-ironstone, with scattered very small plant fragments	0.2
Mudstone, shaly, brown, with some small plant fragments	0.15
Clay-ironstone	0.3
Silty clay, greyish fawn, with plant fragments, passing down to less silty clay	1.5
Sandstone, pale fawn, massive, passing down to	1.8
Silts, white, passing down to	0.9
Clay, pale grey	0.3

About 1.2 m of massive silty fine-grained sandstone at the top of the Ashdown Beds was exposed in the sides of the sunken lane [5588 1840] 100 m west of Lions Green. A thin Top Ashdown Pebble Bed was traced by augering on the hill-slope south of this section.

A deep gully on the north-east side of the road [5462 1594] north-north-west of Holdens Farm exposed the following section:

	Thickness m
WADHURST CLAY	
Silty clay, fawn and pale grey	0.6
Sandstone, coarse-grained, ferruginous (Top Ashdown Pebble Bed)	0.05
ASHDOWN BEDS	
Sandstone, fine-grained, white to cream, massive, passing down to	3.05
Silts, cream to white	1.5

Up to 1.8 m of cream fine-grained sandstone with traces of festoon-bedding were exposed in an old pit [5536 1629] on the east side of the road 450 m north of Stream Farm.   BY

The Heathfield No. 7 Borehole [5859 2149] (Falcon and Kent, 1960) proved 109 m of Ashdown Beds above Purbeck strata. Alternations of sandstones and mudstones were present to 40 m depth, below which siltstones and mudstones dominated the sequence. Ostracods were recorded at 67 and 91 m. The Heathfield Station Well [5804 2135] proved 103 m of Ashdown Beds above the Purbeck Beds.   RDL

An exposure in an old sand-pit [5870 2129], 140 m SW of Gibraltar Tower near Heathfield, showed:

	Thickness m
Sandstone, fine-grained, pale yellowish grey, rubbly	2.4
Sandstone, pale yellowish grey, massive, cross-bedded	1.5

The massive sandstone forms a good feature in this area and is about 100 m above the base of the Ashdown Beds.

In the stream 500 m ENE of Sandy Cross 4.7 m of alternating silts and fine-grained sands are exposed [5885 2050]; the sands are cross-bedded locally.   RABB

An old sand-pit [5875 1963] 650 m NW of Sapperton Farm provided the following section:

	Thickness m
Sand, medium- to coarse-grained, fawn	0.9
Clay, pale grey and brown, with sandy partings	0.05
Sand, medium- to coarse-grained, with small dark brown sandstone pellets in top 0.03 to 0.05 m	0.1
Clay, pale grey and brown, with sandy partings	0.08
Sand, medium- to coarse-grained, pale fawn, with small sandstone pellets in top 0.03 m	0.3

At Maynard's Green [5815 1879] 1.2 m of white silty sand with local cemented layers and traces of cross-bedding were exposed at the top of the road cutting on the west side of the B2203 road. In the east bank of the stream 170 m upstream from the bridge [5863 1852] at the eastern end of Sicklehatch Lane, Maynard's Green, the following section was recorded:

	Thickness m
Sandstone, fine-grained, white	0.6
Clay, mottled orange and grey	0.9
Sandstone, fine-grained, white	0.15
Clays, shaly, brownish grey, with thin sand partings and small plant fragments	0.3
Siltstone, pale grey, laminated	0.75
Clay, pale grey	0.3

The beds here dip south at 5°. A clay bed, 0.15 m thick, within a similar sequence exposed 4.6 m downstream from this section contains abundant fresh sphaerosiderite.

The disused railway line at Horam [5810 1710] cuts through the strong northward-facing scarp formed by the Top Ashdown Sandstone. One section [5779 1727], 1.8 m high, showed poorly laminated pale grey siltstone.

An exposure [5597 1513] in the west side of the road 650 m SSW of Stonehill Farm showed 0.3 m of grey, hard, massive, fine-grained sandstone, overlain by 0.6 m of dark brown sphaerosiderite rock, and by 0.6 m of pale fawn silt. At Pick Hill [5674 1544] a section showed 1.2 m of thinly bedded, silty, fine-grained sandstone with traces of low-angle cross-bedding.   BY

The East Hoathly Borehole [5186 1603] proved 22 m of Ashdown Beds below the Wadhurst Clay. The Top Ashdown Sandstone was 5.5 m thick and capped with a thin pebble bed (Lake and Young, 1978). A sandstone 15.5 m below the top of the formation had an eroded top and became finer downwards; this unit has been correlated with that recorded at a similar level in the Glynleigh Borehole [6085 0637]. Estheriids were recorded in adjacent beds at both localities.   RDL

The road 300 m SW of Brookside cuts through the scarp formed by the Top Ashdown Sandstone and the south bank of this cutting [5890 1704] showed the following:

	Thickness m
WADHURST CLAY	
Clay, grey	1.2
ASHDOWN BEDS	
Sand, medium-grained, ferruginous	0.15
Silty sandstone, fine-grained, cream to white	1.2

On the east side of the road [5931 1767] 210 m SSW of the Brewers' Arms public house about 1.5 m of white silts and fine-grained sands with festoon-bedding were seen.

Exposures of sandstones with some clay beds, in places disturbed by valley bulging, are common in the banks of the stream between Caller's Corner [6109 1949] and Beckington Bridge [6024 1865]. A section in the east bank of the stream [6073 1895] 570 m NE of the bridge showed the following beds:

	Thickness m
Sand, pale fawn, massive	0.15
Clays and silts, laminated	0.6
Sandstone, fine-grained, pale fawn, with clay laminae and cross-bedding in lowest part	0.6
Clay, pale grey, laminated, with fine-grained sandstone beds and lenses (channel fillings) up to 0.3 m thick	0.75
Sandstone, pale fawn, massive with some festoon-bedding	0.9

The south-eastern bank of the stream [6114 1957] immediately opposite Caller's Corner showed:

	Thickness m
Sandstone, fine- to medium-grained, yellowish brown, festoon-bedded	1.2
Clay, pale grey to brown, with a persistent 12-mm lignite parting	0.08
Thinly alternating fine- to medium-grained (locally coarse-grained) micaceous sandstones, with local carbonaceous smears, in beds up to 0.08 m, and pale grey laminated silty clays in beds up to 0.05 m thick	1.0

The relatively steep dip (up to 22° NE) of these beds is probably the result of valley bulging. A small section in the east bank of the lane [6111 1945] 60 m south of Caller's Corner showed thin lignite partings up to 10 mm thick in brown coarse-grained sandstone. On the west side of the lane [6111 1934] farther south the following beds were exposed:

	Thickness m
Silts, pale fawn with some red mottling	0.4
Silty clay, pale brownish grey	0.15
Sandstone, fine-grained, fawn, on channelled surface	0.45
Silty clay, pale grey to brown	0.75

In the west bank of the road [6077 1568] 350 m south of Cralle Place about 0.6 m of massive cream fine-grained sandstone at the top of the Ashdown Beds is overlain by up to 1.2 m of clay and silty clay with local sandstone beds up to 0.05 m thick, probably belonging to the Wadhurst Clay. No pebble bed was detected. A section [6262 1647] in Clippenham Stream 590 m ESE of Beech Hill showed 2.7 m of well-bedded cream siltstones and clayey siltstones cut by a minor NNE–SSW-trending fault. At the lane junction at Kingsley Hill [6157 1801] a section showed 0.6 m of greyish fawn silty clay, probably basal Wadhurst Clay, resting on a rippled surface of massive pale fawn sandstone of which 1.5 m were exposed.

In the stream [6231 1948] 400 m east of Beacon Land up to 0.9 m of thickly bedded, pale fawn, fine-grained sandstone was observed. A similar sandstone, probably part of the same unit, was also seen in an exposure on the east side of the road at Little Marklye [6268 1938] and in the small stream 50 m north of this [6270 1943]. Similar pale fawn massive sandstone was present in exposures, up to 0.6 m high, in the east bank of the road between Rushlake Green [6270 1845] and Marklye [6260 1876].

North-north-east of Rushlake Green, on the west side of the road to Turner's Green, a section at Forster's Farm [6301 1910] showed pale fawn, fine-grained sand becoming finer-grained downwards, 1.2 m, overlain by greyish brown and purplish brown silty clay, 0.38 m, and pale fawn silts, 0.9 m. Up to 2.4 m of pale fawn silts were exposed in the degraded east bank [6297 1872] of the road about 200 m north of Stone House, Rushlake Green; they contained a brown ferruginous hard sandstone, 0.23 m thick, 1.2 m from the top of the section.

South-south-east of Rushlake Green old sand-pits at the southern end of Great Iwood [6320 1760] are now largely obscured. A section at the western end of the workings showed up to 1.2 m of pale fawn, faintly cross-bedded, relatively coarse-grained sand with clay pellets and two distinct pellet beds. The lower pellet bed was 50 mm thick and showed abundant clay and lignite flakes in a clay matrix. The upper bed, 25 mm thick and 0.44 m above, contained small lignite fragments.

An old pit at Sandbank Farm [6356 1845] showed 0.6 m of pale fawn massive coarse-grained sandstone and a similar thickness of dark brown massive medium-grained sandstone with abundant weathered sphaerosiderite was exposed in an old pit on the east side of Grovelye Lane [6468 1820] 150 m ENE of Grovelye Farm.

The following section was recorded in the north bank of Christians River, 380 m SE of Dean Farm [6441 1730]:

	Thickness m
Sand, medium- to coarse-grained, yellowish brown with carbonaceous streaks	0.45
Clay-ironstone, pale fawn	0.05
Sand, medium- to coarse-grained, yellowish brown with carbonaceous streaks	0.15
Clay-ironstone, pale fawn	0.05
Clayey sand, medium- to coarse-grained, with lignitic fragments	0.03
Clay-ironstone, pale fawn	0.03
Alternating beds of medium- to coarse-grained, yellowish brown sands with carbonaceous streaks up to 0.08 m thick and pale fawnish grey clays up to 0.15 m thick	0.45

BY

Beside Willingford Stream [6527 2053], on the western side of Brightling Down, 2.4 m of yellowish brown flaggy sandstones and silts were seen to be strongly folded, probably because of valley bulging: in places the bedding planes are vertical. Just upstream from the latter locality an old quarry [6538 2030] showed:

	Thickness m
Sandstone, yellowish brown, with plant remains	0.6
Sandstone, ferruginous	0.08
Sandstone, yellowish brown, with silty partings up to 0.08 m thick; ripple-marks; plant remains including rootlets apparently *in situ*	1.4
Sandstone, pale yellowish grey, massive	2.4

The section is about 60 m above the base of the Ashdown Beds.

The many small quarries in the Brightling Down area, all dug for sand or sandstone, indicate the sandy nature of the lower part of the Ashdown Beds.

An exposure [6504 1854] south-west of Dallington church showed 1.2 m of massive quartzose sandstone (Top Ashdown Sandstone) overlain by 0.6 m of flaggy ripple-marked and cross-bedded siltstones with medium grey mudstones (Wadhurst Clay). No pebble bed was seen at the base of the Wadhurst Clay. In a road section [6539 1835] about 390 m ESE of the last exposure, 4.6 m of the Top Ashdown Sandstone were exposed. Again no pebble bed was recognised. The massive sandstone was estimated to be about 7.6 m thick.

Around Dallington the Top Ashdown Sandstone makes a striking feature and small quarries expose up to 2.4 m of massive quartzose sandstone [6566 1889]. East-south-east of Dallington church a stream section [6607 1894] showed:

	Thickness m
WADHURST CLAY	
Silty clay, pale grey, with thin siltstones	0.6
Sandstone, pale grey, ripple-marked	0.05
Pebble bed	0.025
ASHDOWN BEDS	
Sandstone, fine-grained, yellowish brown, flaggy	0.9

A road section [6601 1800] 570 m west of Herring's Farm exposed 0.3 m of a massive yellowish brown sandstone (Ashdown Beds) overlain by about 1.8 m of thinly interbedded silts and medium grey clays (Wadhurst Clay). No pebble bed was found.

In the stream through Wheeler's Wood [6694 1715] three massive sandstones, each about 3 m thick, are separated by pale grey silts. Downstream there are excellent sections in deep ravines. About 590 m ESE of Little Redpale Farm cliffs have been cut [6649 1669] in massive pale grey and yellow sandstones near the top of the Ashdown Beds. A stream section through Pagden Wood [6621 1684] 260 m ESE of Little Redpale Farm, showed 5.4 m of alternating pale brown sandstones and silty sandstones in the upper part of the formation. Similar sections are present all along this stream.

In the stream [6553 1639] west of Bucksteep Manor there are many exposures in the uppermost 25 m of the Ashdown Beds. There is also a road cutting [6545 1515] through the Top Ashdown Sandstone south-west of Peartree Farm. RABB

On the valley-slopes of the Nunningham Stream (also called Trulilows Stream and Pebsham Stream) near Bodle Street Green a composite feature has developed at the level of the Top Ashdown Sandstone. Up to 1.2 m of thickly bedded sandstone, with silty partings in the upper beds, were exposed in the lane-side [6465 1470] near Pebsham Shaw, just below the outcrop of the base of the Wadhurst Clay.

No distinct Top Ashdown Pebble Bed was proved by augering in the Nunningham Stream valley. The junction of the Ashdown Beds and Wadhurst Clay is apparently gradational in this area.

A bank exposure [6699 1358] at Henley's Hill in the Hugletts Stream valley showed:

	Thickness m
WADHURST CLAY	
Silts, ochreous, pale grey, platy	0.5

	Thickness m
Clays and silty sandstones alternating	0.15

ASHDOWN BEDS
Sands and clays alternating at about 4-cm intervals.

The sand units are lenticular, with rippled tops	0.36
Sandstone, medium-grained, off-white, friable	0.66
Silt, ochreous, pale grey	—

Fragments of sideritised shelly limestone were noted in the hillwash above the section.

A roadside section [6523 1500] north of Bodle Street Green exposed:

	Thickness m
Talus: silty hillwash	0.3

WADHURST CLAY

Silts and sands interbedded, ochreous	0.05

ASHDOWN BEDS

Sandstone, fine-grained, ochreous and white mottled, thickly bedded with purplish brown silt partings	1.2
Clayey silts, purple and pale grey, with thin sandstone lenses. Slightly rippled top	0.25
Sandstone, very fine-grained, pale grey	0.3

A stream section in Lower Gill [6561 1436] showed 1.5 m of ochreous pale grey, thinly bedded, very fine-grained sandstone, at the Top Ashdown Sandstone horizon.

The BGS borehole at Glynleigh [6085 0637], in the south-east of the district, showed a transitional passage from Ashdown Beds to Wadhurst Clay, with no pebble bed (Lake and Young, 1978). The Top Ashdown Sandstone here was 3.4 m thick.   RDL

Wadhurst Clay

To the south of Newick there is a narrow fault-bounded inlier of the Wadhurst Clay largely obscured by Head. Outcrops occur in the Oxbottom area [409 204]. Old pits exist in Mill Wood [4117 2047]. The drift-obscured faulted contact between the Wadhurst Clay and the Ardingly Sandstone can be traced hereabouts by a line of springs. Farther east clay can also be augered in Broomy Shaw [4330 2050]. Springs to the east, in Sharp's Hanger [4370 2055], allow the faulted Wadhurst Clay/Ardingly Sandstone contact to be mapped.

A borehole at the Uckfield Laundry [4786 2068] proved a complete thickness of 55.5 m of Wadhurst Clay, whereas another at the Grammar School [about 4725 2135] proved 57 m. An incomplete thickness of 40 m beneath 6.1 m of Head, and resting on the Ashdown Beds, was drilled in the Uckfield Waterworks borehole [4820 2176].

White (1926, p.20) recorded the Wadhurst Clay in the railway cutting [about 4885 2182] near Hempstead as 'shale with layers of Tilgate Stone'. The lower boundary is not seen but the upper boundary can be traced in the area to the south by augering the red clay beneath the Lower Tunbridge Wells Sand, by springs, and by the numerous old pits which were opened in the uppermost clay.

In the Framfield area a maximum of 13 m of the basal Wadhurst Clay rests on the Top Ashdown Pebble Bed (p. 24).

There is a small outcrop of the uppermost Wadhurst Clay around Bunce's Pit [4360 1980]. Bell-pits in Lodge Wood can be found over a 400-m tract [4420 1976 to 4460 1982]. They presumably indicate proximity to the base of the Wadhurst Clay. Red clay close to the contact with the overlying Lower Tunbridge Wells Sand has been noted in the Homestead area [4490 1895]. Red clays have also been observed at somewhat lower levels [4514 1913; 4554 1886; 4560 1867] are an estimated 10 to 15 m above the base of the formation.

Benjamin Ware's Tile Works [481 190] closed in July 1970; in September 1970 degraded sections showed up to 3 m of grey shaly clay with ironstone nodules and siltstone lenses up to 0.45 m thick. Ostracods were common on the lower surface of bottom casts of the siltstones. At one place [4813 1933] large slabs of 'Paludina' limestone up to 75 mm thick lay on the floor of the pit. Elsewhere scattered fragments, 10 mm thick, of limestone composed of crushed shells of Neomiodon sp. were observed. The general dip was 12° at 170°.

A well at the Barley Mow Inn [4880 1922] started at 47.9 m OD, an estimated 8 m below the base of the Lower Tunbridge Wells Sand, and proved 33.5 m of Wadhurst Clay on Ashdown Beds.

Outcrops of Wadhurst Clay along the unnamed stream and its tributaries to the south of Little Horsted are largely covered by Head. The Head typically obscures all the lower slope below the Wadhurst Clay/Lower Tunbridge Wells Sand junction down to the valley bottom. The boundary between these two formations can locally be traced beneath the Head by the springs issuing from the contact. In a relatively drift-free tract west-north-west of Peckhams the outcrop is marked by a string of large degraded pits (Limes Pit [4775 1725], Boye's Pit [4820 1720], Whitelocks Pit [4880 1710] and Weavers Pit [4940 1700].   CRB

The red clay at the top of the Wadhurst Clay is easily traced along the north-facing scarp capped by the Tunbridge Wells Sand between Honey's Green [5025 1755] and Annan [5090 1765] and around Barham House [5310 1715]. Augering on the scarp face east of Annan suggests that here up to 1.2 m of grey clay lie between the red clay and the basal sands and silts of the Tunbridge Wells Sand.

Old bell-pits are common in the woods east of Crouch's Farm [5299 1799] and in Hawkhurst Common Wood [5300 1900]. The basal beds of the Wadhurst Clay, consisting of clays, silty clays, siltstones and some shelly ironstones, were exposed in sections on the north side of the lane south of Hawkhurst Common Wood [5323 1857] and in the small stream [5425 1793] 230 m ESE of Crouch's Farm.

Bell-pits are numerous near the base of the formation in the southern part of Summersbrook Wood [5490 1665], known locally as 'Minepit Wood'.

Large shallow pits in the basal Wadhurst Clay around Lions Green [5593 1842] were presumably worked for ironstone. The local name 'Furnace Wood' for the woodland [5700 1800] 400 m NW of Sharp's Corner, recalls the former iron industry. There are numerous bell-pits near the base of the formation in this woodland and there is evidence of a smelting site [5665 1797] near the house known as 'The Furnace', 750 m WNW of Sharp's Corner. Iron slag is abundant in the soil here. The sites of former 'hammer ponds' can be seen along the stream immediately to the west.

The degraded pit [5715 1776] south of the road from Lions Green to Sharp's Corner exposed 1.5 m of red clay below the base of the Tunbridge Wells Sand. Old ironstone workings are common at the base of the Wadhurst Clay in the fields and woodland [5740 1708] about 300 m SW of Horam Manor. Iron slag is plentiful in the field [5764 1589] 300 m NNE of Gamelands, near the cottage known as 'Cindergill'.

The Horam Brickworks pit at Marle Green [5905 1595] provided the most extensive exposures of Wadhurst Clay in the district. The northern face of the pit showed the following disturbed section in 1971:

	Thickness m
Clay, shaly, pale grey	0.6
Siltstone, pale grey, finely laminated, cross-bedded with abundant channels and load structures; some rootlet traces near top and abundant Neomiodon sp. shells at base	0.45
Clay, shaly, pale grey; Equisetites sp. stems common as flattened impressions, some as solid casts, some stems erect and in growth position; occasional thin shell-debris laminae; channelled top	0.9

The section in the eastern face was as follows:

	Thickness m
Shaly clay, grey and fawn weathering	0.9
'Cyrena' limestone, ferruginous	0.15 to 0.2
Mudstones, shaly and inter-laminated with fine siltstones, passing down to	0.3
Clays and shaly clays, grey and fawn, laminated, numerous thin shell-debris laminae and scattered *Equisetites sp.* impressions	1.2
'Cyrena' limestone, ferruginous	0.15 to 0.2
Siltstones, grey and fawn, laminated, with some load structures (especially near the top)	0.6
'Cyrena' limestone, ferruginous, thin in south-east of pit, wedges out northwards	0 to 0.05
Clay, pale grey, laminated, with occasional bands of tabular clay-ironstone nodules, numerous shell-debris laminae, numerous ostracod-rich laminae, some *Equisetites sp.* impressions; a large reptilian bone fragment and fish scales near top, beneath the channelled upper surface	1.8

These two sections were separated by several metres of unexposed ground but it is estimated that the top of the first corresponds roughly with the bottom of the second. The floor of the pit is about 6 m above the base of the Wadhurst Clay.

A section on the south side of the road 260 m NE of Cowden Hall [5978 1620] showed the basal beds of the formation, as follows:

	Thickness m
Clay, shaly, pale brown and grey, passing down to	0.6
Clay, silty, with fine-grained sandstone beds up to 2.5 cm thick	0.1
Sandstone, cream, fine-grained, massive,	1.0
Clay, silty, greyish brown, with medium-grained sandstone beds up to 4 cm thick	0.3

A stream [5906 1784] 340 m west of the Brewer's Arms public house at Vine's Cross exposed about 0.6 m of grey shaly mudstones with some siltstone beds up to 5 cm thick, thrown into a sharp anticlinal fold by valley bulging.

Blocks of clay-ironstone are plentiful in the overgrown pits [5995 1850] near the base of the Wadhurst Clay in the woodland known as Brick Kiln Shaw, 300 m NE of Hale Hill. Large quantities of black glassy slag occur in the fields on the opposite side of the stream [6010 1865] 550 m NE of Hale Hill. An area of alluvium upstream from the remains of a dam [6000 1860] marks the site of a former hammer pond (Straker, 1931).

Other old ironstone pits, many of them flooded, are common in the woodland south of Warbleton [6090 1785] and slag fragments occur in the field [6045 1745] 450 m WNW of Tilement Farm. Old dams and silted-up hammer ponds can be distinguished in the valley west of Tilement Farm [6030 1754; 6043 1699]. The local name 'Steelforge Wood' [6060 1695], 450 m SW of Tilement Farm, recalls a former smelting site (Straker, 1931).

The following sequence was exposed in 1971 in the west bank of the road [6077 1567] 350 m south of Cralle Place:

	Thickness m
WADHURST CLAY	
Clay and silty clay with at least one 5-cm medium- to coarse-grained sandstone bed containing thick mud flakes up to 1 cm across	1.2
ASHDOWN BEDS	
Sandstone, cream, fine-grained, massive	0.6

Allen (1947, p.313) recorded an exposure of the Brede *Equisetites lyelli* Soil Bed at this locality, but it is no longer visible. Old pits, presumably dug for ironstone, are numerous in the fields immediately to the west.   BY

A sandstone crops out on the valley sides to the west of Springham Wood [592 143]. A stream section [5882 1429] exposed 0.9 m of thinly bedded sandstone, and up to 1.2 m of similar beds are present in the stream banks [5929 1394] north of Sandrock Wood. The latter section shows the silt content increasing downwards by interlamination. Evidence from field-brash suggests that a bed of 'Tilgate Stone' occurs in association with the sandstone, which is apparently in the higher part of the Wadhurst Clay.

In the area between Trulilows [629 148] and Chilsham [635 136] at least two prominent ironstones have given rise to topographic features and associated old workings.   RDL

Numerous bell-pits can be seen in the woods [6300 1715] 450 m NW of Iwood Place Farm. The woodland [6355 1750] north of the cross-roads 600 m NNE of this farm contains a number of flooded old ironstone pits.   BY

An old clay-pit [6687 1970] in the small faulted outlier at Hook's Farm showed sections in the lower part of the Wadhurst Clay. In the top of the southern end of the pit was fine-grained sandstone representing the Ashdown Beds on the other side of an E–W-trending fault with a throw of about 12 m down to the north. The Wadhurst Clay sections, although very weathered, showed:

	Thickness m
Mudstones, medium grey, with thin siltstones	0.6
Clay-ironstone, nodular, weathering reddish brown	0.025
Mudstones, medium grey, with thin siltstones weathering green in places; load casts; *Neomiodon sp.*	0.45
Mudstones, medium grey, with thin pale grey siltstones; several beds of clay-ironstone nodules; patchy red and green colour in places	0.9

The outlier at Dallington is faulted along its southern margin; the fault trends approximately E–W and downthrows up to 18 m to the north. A section south of High Wood [6753 1842] showed 0.6 m of medium grey mudstones and siltstones with *Equisetites lyelli* rootlets; also a 25-mm-thick nodular clay-ironstone and several thin limestones with *Neomiodon sp.* A little way below is a massive sandstone, the Top Ashdown Sandstone. The *E. lyelli* rootlets in growth positions indicate a soil possibly equivalent to the Brede Soil Bed (Allen, 1947). Loose siltstones, commonly with load casts, are present in the soil above the lower part of the Wadhurst Clay, indicating the silty nature of this part of the sequence. Bell-pits for ironstone are very common, the most spectacular being at Grip Hill Shaw [6530 1835] near Stream Farm, in Haselden Wood [6732 1903; 6775 1845], in a wood near Thornden Farm [6765 1708], and in Reed Wood [6720 1540], about 700 m from Ponts Green.   RABB

Sandstone debris on the slopes of the Frankwell Gill valley [6723 1425] west of Bray's Hill suggests that there is a sandstone about 12 m below the Tunbridge Wells Sand.

Allen (1947) recorded the following localities which exposed soil beds within the Wadhurst Clay of the Lewes district. Many of these sections are now degraded.

1 Old sand-pit on west side of Church Stile Lane 0.54 km NE of Framfield.
2 West road-bank 0.4 km south of Cralle Farm, Warbleton.
3 Sand-quarry east of Hoad's Wood, 1.2 km SW of Dallington church.
4 Southern lane-bank at Sandhill Cottages, 2.4 km NNW of church at Bodle Street Green.
5 Old sand-pit east of lane 0.4 km SE of No. 3.
6 Old quarry in north-east angle of crossroads, Redpale Farm, 2 km NE of Bodle Street Green.
7 Eastern lane-bank north of Nunningham Stream, north-east of church at Bodle Street Green.

8   Western lane-bank south of Christian's River, 0.8 km north of church at Bodle Street Green.
9   Northern lane-bank 100 m west of crossroads at Ponts Green.
10 Old sand-pit east of lane 1.6 km NE of Windmill Hill.
11 East lane-bank 0.8 km south of Bray's Hill, 0.8 km NE of No.10.   RDL

Lower Tunbridge Wells Sand

Crags of Ardingly Sandstone can be seen near Rock Hall [4757 2180] close to the A22 road. A nearly continuous line of rocky crags can be traced for 1.5 km in a west-south-westerly direction on the west side of the A22 road, from Uplands [4740 2215] to Combe Banks [4700 2190] and thence to beyond The Rocks [4670 2160]. Dips in this region range from 2 to 5° NW. Crags also occur low down in the valley of the stream which flows through Shortbridge [4510 2125] between the bridge and a place [4620 2200] 1.35 km NE. A dip of 10°ESE was noted [4590 2175] and may be on one limb of a syncline associated with the nearby fault.

In the sunken road [4400 2065] south of Sharpsbridge a massive sandstone weathering flaggy was seen; the dip is 10°SE. Rocky crags [4446 2046] were noted at alluvium level in Rocky Wood and low crags [4605 2119] protrude through the soil cover 130 m north of Copwood. The Ardingly Sandstone forms a line of low crags south of the river through the southern part of Uckfield; dips in this area range from 5°SW to 10°S and indicate that the outcrop here forms the southern limb of a shallow anticline with a WSW–ENE axis passing through the centre of Uckfield. Professor P. Allen (personal communication) noted the Top Lower Tunbridge Wells Pebble Bed in a roadside exposure [4728 2076] 130 m south of Uckfield railway station. The most easterly exposure [4739 2087] of Ardingly Sandstone noted in this area is 150 m east of the station. The total thickness of the Lower Tunbridge Wells Sand as proved by a well [4786 2068] at the Laundry in Uckfield is 32 m, of which about 12 m can be assigned to the Ardingly Sandstone.

In the area east of New Town and north of Framfield some 15 m of beds in the lower part of the Lower Tunbridge Wells Sand crop out. They consist mostly of fine brown and buff sands and sandstones. Thin flaggy sandstones were noted locally in some of the sunken lanes [4925 2090; 4938 2064] and a more massive sandstone, weathering flaggy, was seen in an old quarry [4890 2050]. The dip in the quarry was 5°N, but the regional dip for this outcrop appears to be south-westward.

At Ridgewood Hill [475 198] beds of the lower division dip at 5° to 10° NNE towards a fault. Exposures in several old pits [4730 1990] and road cuts showed up to 4.5 m of thickly bedded fine-grained sandstone which weathers flaggy. Although lithologically similar to the Ardingly Sandstone, these beds appear to be lower in the succession.

The Ardingly Sandstone crops out in the base of the disused railway cutting [4068 2000], 1.5 km SW of Newick, where it was formerly well-exposed, overlain by the Grinstead Clay.

In the road sections [409 203] to the south of Oxbottom there are good exposures of festoon-bedded sandstone. A dip here is 6° S. Similar exposures in sunken lanes to the south of Founthill [4205 2013] are in massive or festoon-bedded horizontal sandstones. An anomalous dip of 40°N [4210 2012] is attributed to cambering.

A section [419 201] measured along the road 200 m WSW of Founthill (Plate 3) showed:

	Thickness m
Sandstone, thin, flaggy, with a channelled surface	1.2
Sandstone, fine-grained, massive, cross-bedded	1.5
Sandstone, thick, flaggy, locally passing into massive sandstone	5.7
Sandstone, massive, friable	2.7

East of Founthill a break of slope was taken to mark the contact of the Ardingly Sandstone with the lower part of the Tunbridge Wells Sand. Massive buff fine-grained sandrock weathering flaggy was noted in the sunken lane [4165 1966] 150 m NNE of Ridgeland Farm; the dip in this vicinity is 4° to 6° SSW.

In Newick Park 1.5 m of massive fine-grained sandstone dips at 6° towards the Longford Stream valley. Farther north, road cuttings [4235 1975] exposed only soft sandstone, but 200 m NW massive sandrock again crops out. Sandrock (Ardingly Sandstone) has been worked from beneath the Grinstead Clay in a small pit [4226 1916] in Newick Park and also in a pit [4260 1996] 600 m WNW of Broomlye Farm. The advantages of this method of extraction were that the rock could be easily worked and it then hardened after exposure. Mantell (1833, p.206) stated that 'Coal of the Bovey kind' was found in Newick Park and excavated to a depth of 0.28 m.

Crags of massive sandrock (Ardingly Sandstone) occur at the roadside [4322 1906] east of Gipp's Farm, where the beds dip at 4°SE, and in the wood to the west [4312 1906].

An old pit [4370 1874] 350 m ESE of Sutton Hall exposed 2.1 m of massive cross-bedded buff fine-grained sandstone. At the top of one crag were seen 150 mm of coarse sand with grains up to 3 mm in diameter; the dip was 7°SW. In the river cliff [4400 1825] on the west side of the River Ouse 850 m SE of Sutton Hall crags up to 2.1 m high of massive sandstone are exposed.

In Isfield the Ardingly Sandstone has been extracted from a number of abandoned pits in which exposures up to 5 m deep are still visible. One such exposure [4472 1789], on the west side of the road, showed 1.2 m of massive buff fine-grained sandstone overlain by 1.2 m of dark grey shaly clay of the Grinstead Clay; the dip here was 13°SW; no pebble bed was noted. In the quarry [4485 1785] on the opposite side of the road an exposure at a slightly lower level was in 1.4 m of massive fine-grained sandstone beneath 0.9 m of soft flaggy fine-grained sandstone; the dip was 5°WSW. Additional exposures 230 m to the east showed dips of 3°NNW and 2°SSW. At one of these localities [4508 1785] 4.6 m of massive white sandrock at the base of the section had large channel scoops on its upper surface infilled with black fine-grained sand up to 115 mm thick; this was overlain by 0.45 m of cream silt.

Farther east, around Little Horsted, the road cutting in the lower beds of the Lower Tunbridge Wells Sand was the type locality of Mantell's (1833, p.191) Horsted Sand, subsequently included in the Lower Tunbridge Wells Sand. A section [4683 1816] measured in the south-western part of the cutting in 1970 showed the following strata dipping at 6°SSW:

	Thickness m
Sandstone, fine-grained, buff	0.2
Sand, laminated, buff and yellow, with grey silt resting on a rippled surface	0.2
Sandstone, fine-grained, yellow, with a ferruginous base	0.15
Sandstone, fine-grained, buff, iron-stained	0.17
Sand, buff, laminated with grey silt	0.03
Sandstone, fine-grained, buff	0.1
Sand, fine-grained, buff, laminated with grey silt	0.1
Sandstone, fine-grained, massive, buff	seen to 0.3

Some 200 m to the north-east, ribs of massive sandstone, 0.6 to 0.9 m thick, protruded through the soil cover of the road bank; the dip was 10°SSW.

A small outlier of the lower part of the Lower Tunbridge Wells Sand is present 800 m SE of Little Horsted; poor exposures show soft, buff fine-grained sandstones.

The two small clay outliers in the Crockstead Green area [494 175] are thought to be Grinstead Clay (p. 30). The Ardingly Sandstone has not been differentiated in this area, although massive sandstones are present below the clay. A break of slope some 5 m above the base of the Lower Tunbridge Wells Sand may represent the base of the Ardingly Sandstone. If this interpretation is correct then this is the farthest east within the district that the Tunbridge Wells Sand can be divided with certainty into its three formations.

A section [4932 1775] in the lane to Crockstead Farm, in about the middle of the formation, showed 2.7 m of white poorly cemented fine-grained sandstone. The sandstone consisted of tabular beds about 0.1 m thick, separated by partings of buff fine-grained sandstone 13 to 25 mm thick; the dip here was 2°SSW. The lane [4965 1742] south-west of Lower Sandhill exposed horizontal yellow poorly cemented massive fine-grained sandstone weathering into units 10 mm thick with honeycomb structure. A borehole [4958 1719] 500 m SSW of Lower Sandhill penetrated 5.5 m of Tunbridge Wells Sand before entering Wadhurst Clay.   CRB

Grinstead Clay

The Grinstead Clay north of Uckfield has a thickness which probably ranges between 5 and 10 m. Its outcrop is marked by a series of old pits. To the west of Downland Farm the 500-m-wide pit-pocked belt of Grinstead Clay [468 221] is a dip-slope outcrop, the dip being about 3°WNW; red clays were noted in the upper beds.

At the eastern limit of the formation an incomplete thickness of 5.3 m was proved in the Laundry well [4786 2068] in Uckfield. The junction with the Top Lower Tunbridge Wells Pebble Bed was noted in a roadside exposure [4728 2076] on the A22 road. East of this locality the formation is a red or mottled red and grey clay. Red clay was also noticed immediately below the base of the Upper Tunbridge Wells Sand close to the railway line [4663 2069; 4669 2078] south-west of Uckfield where the dip was 5°WSW.

In the Broomlye Farm area [430 200], an outlier of Grinstead Clay, capped by Upper Tunbridge Wells Sand, is separated by a fault from a similar outcrop to the south in Newick Park. Here the Grinstead Clay is an estimated 9 to 10 m thick.

In Newick Park mottled orange or yellow and grey clays were augered. The regional dip is about 4°SSW. From this area the outcrop swings south-eastwards and a narrow belt, some 6 to 8 m thick, was followed through Gipp's Wood [430 189] and the grounds of Sutton Hall.

A roadside section [4328 1898] 160 m SE of Gipp's Farm indicated that thin sandy beds occur within the clay:

	Thickness m
Sand, fine-grained, mottled orange and grey	0.7
Clay, silty and sandy, yellowish brown	0.9
Sandstone, fine-grained, brown	0.3
Siltstone, grey	0.45
Sandstone, fine-grained, grey, ferruginous, flaggy at bottom	0.4
Clay and silty clay, red and greenish grey	seen to 2.4

Mottled orange and grey silty clay above this section was augered for 80 m to the south.

East of the River Ouse the thickness of the Grinstead Clay diminishes and appears not to exceed 5 m, and the formation becomes more silty. The most easterly occurrence mapped with certainty is to the north of Wicklands, where grey, or mottled yellow and grey, clay was augered. The clay was formerly extracted from a number of small pits [462 180; 4626 1772] in this vicinity. Red clay is present beneath the Upper Tunbridge Wells Sand.

Two outliers of mottled yellow, cream, grey, pink and red clays at Crockstead Green are tentatively included with the Grinstead Clay, although they occur at an anomalously low stratigraphical level, the base of the clays being only about 17 m above the base of the Lower Tunbridge Wells Sand. The more easterly outlier [494 174] consists predominantly of red clay.   CRB

Upper Tunbridge Wells Sand

Exposures of Upper Tunbridge Wells Sand are rare in the Uckfield and Newick areas. Augering revealed a sequence dominated by silts, but clays, silty clays, fine sands and sandstones also occur.

South and west of Uckfield an extensive outcrop of the formation exists in the New Town [473 205] – Buckham Hill [451 205] area. A clay bed was traced for a short distance on each side of Buckham Hill House, and a small outlier [4540 2095] 550 m NE of the house is probably at the same horizon.

A well [4501 2182] at Gazle Slope penetrated 28 m of the Upper Tunbridge Wells Sand before entering the Grinstead Clay. Flooded pits in the vicinity of Brickyard Farm [4495 2197] indicate the presence of clays or silty clays at this level. Thin flaggy sandstones were noted in nearby roadside exposures [4422 2195; 4447 2192; 4456 2180]. At the first of these localities the beds are horizontal.

Thick sandstone weathering flaggy and dipping at 1°N was noted in the lane [4204 2067] 100 m SSE of Ketches. An old pit [4145 2166] to the north-west of Newick showed 1.8 m of thinly bedded sandstones and silts infilling a hollow in 1.8 m of cross-bedded massive sandstone. According to the owner the lower sandstone overlies 'blue clay'.

Exposure is poor in the area between Chailey and Little Horsted. The contact with the overlying Weald Clay is faulted except to the south-west of Ades [400 190]. Generally only silty sands and fine sands and sandstones were proved by augering. Thick flaggy yellow fine-grained sandstone was noted in Cockfield Lane [4096 1922]; this sandstone appears to form a southward facing dip-slope and overlies a bed of mottled orange and grey silty clay, about 7 m thick, which can be followed in an easterly direction for 1.3 km.

A roadside exposure [4323 1923] 130 m NNE of Longford Bridge showed soft flaggy fine-grained sandstone dipping at 8°S. Ribs of massive sandstone crop out in the railway cutting north-north-east of Isfield Station.   CRB

Tunbridge Wells Sand undivided

The basal Tunbridge Wells Sand forms a prominent north-facing scarp running west–east through Honey's Green [5025 1755] and Annan [5090 1765].

Fine-grained sand overlying a red-mottled grey clay bed was traced for about 700 m along a marked low feature [5080 1615] north-west of Halland Park Farm.

A clay within the Tunbridge Wells Sand has been mapped for about 800 m along the downthrow side of an WSW–ENE fault through East Hoathly village [5225 1635]. Augering showed this clay to be pale fawn and grey, locally with strong red mottling. It may be the same clay as that recorded between 3.73 and 7.25 m in the East Hoathly Borehole [5186 1603] (Figure 5) and correlated with the Grinstead Clay (Figure 6).

Fine-grained cream sandstone was exposed in sections up to 1.2 m high in the sides of the lane [5264 1664] 70 m north of Belmont House, East Hoathly.

Up to 1.8 m of silt overlies the red clay at the top of the Wadhurst Clay in an old pit [5275 1769] 300 m west of Hope's Farm. In the faulted block of Tunbridge Wells Sand south of Barham House an old pit [5303 1691] on the east side of the road exposed 1.2 m of yellow silty sand and fine-grained sandstone. Another overgrown pit [5334 1689] 300 m east of this showed up to 1.5 m of yellow fine-grained sandstone. The following section, near the base of the Tunbridge Wells Sand, was exposed in an old pit [5377 1707] on the west side of the road 660 m east of Barham House:

	Thickness m
Sand, fine- to medium-grained, orange, laminated	0.3
Silt, clayey, pale greyish brown, with fine- to medium-grained sand in laminae and lenses up to 1.5 cm thick, the sand laminae becoming more numerous upwards	1.2

	Thickness
	m
Siltstone, white, massive	0.15
Sandstone, fine-grained, mottled grey and orange, massive	0.3

A small pit [5574 1758] on the west side of the road 700 m south of Lions Green exposed up to 1.2 m of white silts. The Tunbridge Wells Sand caps a strong north-facing scarp south of Sharp's Corner [5735 1780] with a long dip-slope to Horam Manor [5757 1733]. In the south face of an old clay-pit [5717 1775] 200 m west of Sharp's Corner 1.2 m of yellow fine-grained silty sands at the base of the Tunbridge Wells Sand overlie red and grey clays at the top of the Wadhurst Clay. East of Horam [5790 1750] the base of the Tunbridge Wells Sand is again marked by a prominent scarp and the small outlier [5940 1835] 500 m north of Vine's Cross [5940 1780] caps a well-marked scarp.

Up to 1.2 m of massive cream fine-grained sandstone, believed to belong to the Tunbridge Wells Sand, were noted in the east side of Hammer Lane [6070 1695] 450 m SSW of Tilement Farm. In the east bank of the lane [6220 1703] 120 m north of Durrant's Farm 0.9 m of massive, cream, fine-grained sandstone is overlain by about 1 m of cross-bedded, yellow, fine-grained sandstone.

Around Sandhill Farm [6431 1644] the Tunbridge Wells Sand caps a strong north-facing scarp. A small pit [6430 1619] on the east side of the lane 230 m south of the farm exposed the following section:

	Thickness
	m
Silt, clayey, pale grey	0.3
Sand, fine- to medium-grained, pale greyish fawn	0.3
Silt, clayey, pale grey	0.3
Sand, fine- to medium-grained, pale fawn, with carbonaceous streaks and fragments up to in 2.5 cm in diameter	0.9
Sand, fine- to medium-grained, cream, with faint cross-bedding picked out by brown ferruginous bands; channelled top	1.2
	BY

Up to 3 m of silts and sands were exposed in an old pit [6582 1555] about 100 m south of Peartree Farm. Near Brigden Hill a similar section [6662 1567] was in yellowish brown to pale grey fine-grained sandstone. The sandstone is quite massive at the base and becomes more silty upwards. RABB

Sandstone possibly equivalent to the Ardingly Sandstone, has been traced from near Chiddingly Place [5413 1437] to The Park [5509 1405], where it is overlain by pale grey silts and clayey silts. In the valley north-west of The Park and northwards to Hale Green the sandstone is overlain by red-mottled clays and pale grey clayey silts. Red clays were formerly worked at a pit [5462 1480] north-east of Parsonage Farm and augering in the road-bank east of this location [5475 1478 to 5485 1482] proved alternations of red clays and greyish green clays, suggesting a localised development of the Grinstead Clay.

East of the main tributary stream to the Cuckmere River, the Ardingly Sandstone-equivalent forms a wide outcrop on the south dip-slope from near Carter's Farm [564 142] to Alland [573 138]. Near Popp's Farm [5700 1390] 1.5 m of thickly bedded to massive fine-grained sandstone showing some cross-bedded units was exposed. In the area just described the sandstone is overlain by ochreous pale grey silt and locally red-mottled pale grey clayey silts. A sandstone approximately 15 m higher in the succession occurs as outliers at Parsonage Farm (where it is overlain by red-mottled silty clays), west of Carter's Farm and east of Hamly Bridge [561 136] (Figure 7). West of Parsonage Farm the sequence is complicated by lenticular (channel) sandstones. At a pit [5380 1495] near Frith's

Farm, up to 1.2 m of a ferruginous pellet bed, with silt and lignite pellets up to 2 mm in diameter in a silty matrix, was seen to cut down into clayey silts (up to 1.2 m) which overlie 0.6 m of massive sandstone.

In the fault-bounded area centred on Gun Hill [565 145] a similar 30-m succession was observed, comprising (in upward order) sandstone; silts and clayey silts, pale grey, locally red-mottled; sandstone; and clayey silts, red-mottled pale grey. The lower sandstone, which is possibly equivalent to the Ardingly Sandstone, was poorly defined. A roadside section near Gunhill Wood [5677 1491] showed the lower, massive sandstone, and small exposures near Gun Hill [5655 1462; 5662 1426] and near Pickley Wood [5715 1459] displayed the higher sandstone unit.

A pit [5411 1290] at Burgh Hill exposed:

	Thickness
	m
Silts, clayey, buff to grey, finely laminated	1.2
Silts, clayey, purple and brown, locally lignitic, with rootlets	0 to 0.2
Sandstone, finely laminated, with a sharp slightly eroded top	0.9

Red-mottled clays are present below the sandstone which crops out at Thunder's Hill [5521 1313] and is apparently equivalent to that last described.

Mantell (1833, p.244) described plant-bearing sand with *?Lycopodites*, *Lonchopteris mantelli* and *Sphenopteris sp.* on the roadside leading from The Dicker (A22) to Chiddingly; the exact locality is not known.

Around Wellshurst [580 146] the lower beds of the Tunbridge Wells Sand consist of sands, silts and clayey silts. A pit [5789 1441] near Wellshurst showed 0.6 m of massive pale grey sandstone overlain by 1.2 m of thinly bedded ochreous sandstone; the dip is to the west. In the valley east of Leyhurst Farm a stream bank section [5893 1495] exposed the following beds in the lowest part of the subgroup, dipping to the north:

	Thickness
	m
Silts, clayey, ochreous pale grey	2.4
Siltstone, pale grey, bioturbated	0 to 0.3
Mudstone, silty, shaly, medium grey, bioturbated	1.1
Silts, pale grey, soft	1.2
Siltstones, thinly bedded, ochreous	0.9

A well at Lealands [5800 1330] proved 29 m of Tunbridge Wells Sand and apparently penetrated 1.5 m of Wadhurst Clay, suggesting that the sandstone which crops out on the west-facing slope below lies approximately at the level of the Ardingly Sandstone. A pit nearby [5823 1344] exposed 2.1 m of cross-bedded massive sandstone, apparently at the same horizon, overlain by clayey silts.

South of the tributary to the Cuckmere which flows from south of Chiddingly to Lea Bridge [5793 1290], and in the Hellingly area, higher beds of the Tunbridge Wells Sand crop out, the beds consisting of silts and clayey silts with subordinate sandstones. Locally the sandstones form mappable units, although they pass laterally into silts. More clayey beds marking the upper transition to Weald Clay are present to the south.

An old pit [5945 1297] near Blackstock Farm showed:

	Thickness
	m
Clays, sandy, ochreous pale grey	0.6
Pellet bed, iron-cemented	up to 0.05
Sandstone, fine- to medium-grained, ochreous pale grey, cross-bedded	up to 2.1

Another pit [5979 1328] exposed the following sequence:

	Thickness m
Silts, clayey, weathered to loam	up to 1.8
Pellet bed (pellets about 1 mm in diameter); dip about 30°	0.6
Gap in sequence	
Sandstone, medium-grained	—

The latter sandstone unit lies at a level approximately equivalent to that of the Ardingly Sandstone; each sandstone is overlain by channel-fill deposits.

In a culvert section [5971 1176] east of Park Farm, 2.5 m of beds above the red-mottled clay group were noted, comprising three fine-grained sandstones 0.15 to 0.18 m thick, alternating with silts; the sandstones have sharp even bases and ripple-marked tops, and are locally ferruginous. The dip is about 15°SE, with downward flexuring increasing southward.

Sands and silts of the lower part of the Tunbridge Wells Sand crop out extensively on the ridge on which Magham Down [610 116] and Herstmonceux [636 125] are situated. A sandstone at an equivalent level to that of the Ardingly Sandstone occurs in the valleys around Deudney's Farm [6167 1200], extending down-dip southwards to near Haffenden's Farm [6222 1124] and as isolated outliers south of Lime Park [637 121] (Figure 7). Beds above this consist of pale grey silts and silty clays, locally red-mottled. A higher sandstone crops out south of Hellingly Hospital [596 124], south-west of Carter's Corner Place [609 124], and as outliers east of Gildridge Farm [612 110], south of Eve's Cottage [6180 1129] and south-west of Upper House [618 117]; at the last locality pebbly sandstone and pellet-bed debris occur in the soil brash.

The Hailsham Borehole [5746 1083] proved 40 m of beds assigned to the upper part of the Tunbridge Wells Sand beneath the Weald Clay (Figure 8).

Silts with sandy partings, 2.4 m thick, were recorded in the road bank [6353 1170] south of Lime Park. Silty clays were formerly dug for brickmaking in pits nearby at Golden Cross [633 110].

In the roadside section near Herstmonceux church [6429 1039] a sandstone 0.4 m thick lies within platy siltstones and is faulted against silts to the south. This minor fault is perhaps allied to the main SW – NE trending fault farther south.

Road cuttings west and north of Wartling showed 1.2 m of grey silts with iron-pan, beneath 1.2 m of well-bedded sandstone [6561 0904], and 1.4 m of ochreous pale grey massive sandstone [6577 0927], respectively.

The fault-bounded salient south of Hailsham apparently lies in the upper part of the Tunbridge Wells Sand on the evidence of a borehole at Rickney [6214 0664] (p.35). Two sandstones within silts and clayey silts are recognised near Sandbanks; the upper one was described by Mr A. H. Taitt (in MS) in an old quarry [586 084] near the cemetery as gently dipping, massive, fine-grained, variably yellow to chocolate in colour and bituminous in part; some 2.4 m were exposed in 1936.

The faulted inlier from Glynleigh to Pevensey lies in the lower part of the sub-group. The Glynleigh clay quarry [607 064] showed the following section:

	Thickness m
Sands, with clay alternations	1.1
Sandstone, with scattered ironstone pebbles at top	0.7
Silt, clayey, pale grey	0.3
Sandstone, silty, carbonaceous	0.2
Silts, pale grey	2.4
Sandstone	0.3
Clay, silty, greenish grey	0.2
Silts, pale grey	0.7
Silts, very carbonaceous (seatearth)	0.6
Silts, pale grey, with thin sandstones	4.6

The Glynleigh Borehole [6085 0637], which was drilled in this quarry, commenced in the lowest silts and continued in predominantly sandy beds to a depth of 7.76 m; a succession of red and green mottled sphaerosiderite-bearing siltstones and mudstones was then encountered, forming a passage series to the Wadhurst Clay (Figure 5), which was reached at 41 m depth (for full details see Lake and Young, 1978). The sequence of mottled silts and clays crops out on the lower slopes around Hankham [619 056]. A borehole near Montague [6265 0595], within the same clay sequence, proved 6.4 m of medium grey silty clays, thus demonstrating the impersistent nature of the colour mottling.

For the most part sands and silts crop out to the west and south of Hankham. Red-mottled clays and silts also occur locally, in the lower part of the valley east of Peelings [618 049]; on the lower slopes below Castle Farm [6385 0478], and along the A259 road through Westham. Spherical pyrite nodules were present in mottled silts in a temporary exposure at Mount Pleasant [6335 0457]. Topley (1875, p.97) also recorded pyrite nodules in the railway cutting 'west of Pevensey' (erroneously ascribing these to the Weald Clay).

Yellow sand with laminae of grey clayey silt was formerly exposed 275 m NW of Westham church [639 047] (White, 1926, p.26) and Mr A. H. Taitt (in MS) recorded 6 m of 'sandy beds varying from white silts to quite coarse ferruginous sandstones' in the cutting opposite Garden Cottages [6279 0452]. Sandstones were formerly worked near Lusteds [6138 0641] and south of Westham [6386 0409].

Bituminous sandstone in the lane south of Chilley [6374 0600] is of a type said by Mantell (1833, pp.172 – 173) to have been used by the Romans for building purposes and the manufacture of projectiles. Similar 'oil sands' have been noted, associated with faults, at Hailsham (Lees and Cox, 1937, p.156), Downash and Rickney (Reeves, 1949, p.266).

Low 'islands' in Hooe Level, east of Middle Bridge, are composed of fine-grained sands and silts of the upper part of the Tunbridge Wells Sand. RDL

Weald Clay

Area west of the River Ouse

In the area between South Street (Chailey) and Barcombe extensive outcrops of Bed 3 (p.22) are present, apparently repeated by strike-faults and affected by minor folds. The marker clay denoting the presence of this group is variably greenish grey to medium grey at surface, with red clay locally at the top. Impersistent sand bodies are present above and within this unit.

In the Chailey brickworks [393 176] the sequence is:

	Thickness m
Head: flinty clay	up to 1.3
Clay, mottled orange and grey, with some red clay	3.0
Mudstone, sideritic	0.2
Clay, mottled orange and grey, and bluish grey	3.0
Sand, fine-grained, yellow	1.1

The sand at the base is equivalent to the highest part of Bed 3. Mottled clays were augered at a lower stratigraphical level in this area. CRB

Fragments of 'Cyrena' limestone were found at the northern end of the degraded railway cutting east of Park Gate [4123 1798], marking the top of Bed 2 (p. 22). Sandstones, showing ripple structures and cross-bedding, were noted farther south in the cutting [4115 1772] within grey mudstones.

Between Balneath Wood [407 175] and Mount Pleasant [424 167], two main sandstones up to 3 m thick are present, and below, between and above them three red clays occur within the grey clays.

The clay marker which is believed to be equivalent to the Large-

'Paludina' limestone (Bed 4, p. 22) was mapped from the western margin of the sheet [391 155] north-west of Winterlands Farm, to near Barcombe Station [4174 1571], where it is covered by drift. A prominent sandstone above these red clays was mapped for some 2.5 km to the west of Barcombe Cross. No limestone brash was encountered in the clay marker, although Large-'Paludina' limestone workings occur near Plumpton Green [364 168] in the adjacent Brighton district (Sheet 318).

The Hamsey brickyard [399 160] formerly exposed beds above Bed 4 to the south of a fault (Reeves, 1958, p.9). In 1973 only a section north of the fault was evident showing:

	Thickness m
Head: silty loam with angular flints	up to 1.5
Silts, clayey, pale grey, weathering to buff and ochreous; clay partings	1.4
Clay, silty, medium grey, with siltier partings	1.8
Clayey silts and silty clays (as above), alternating in 0.2-m units	1.8
Silts, clayey, pale grey, with ironstones in thin clay partings	2.4
Clay, red-mottled greenish grey	1.4

The beds have a slight dip to the north-east with minor small-scale flexuring. The lowest horizon exposed is the highest mottled clay of Bed 3. White (1926) described a section 'by the cross-ways south of Havenstreet Farm' in the 'Hamsey Brick Company's yard' as follows, with thicknesses converted to metric units:

	Thickness m
Soil: loamy clay with flints	—
Clay, silty, mottled grey and brown	3.0
Clay, dark red	6.1
Shale, grey, with lenticles of hard slightly calcareous siltstone and of laminated fine sandstone; *Cypridea valdensis* in the shale	seen to 1.5

This sequence was presumably exposed in the same pit, but because most of the older workings lie south of the fault, this is difficult to relate to the established succession. The description suggests lithologies of Bed 3. Mantell's mention of ironstone septaria with 'Cyrena', Viviparus, fish debris and ostracods (1833, pp.187, 188) at 'Resting-oak Hill' [397 152] is also hard to relate to the sequence, since this location appears to lie above Bed 4. RDL

Area between the River Ouse and Laughton

A thin bed of ironstone was traced for over 1.5 km from north of Rose Hill [4572 1678] to near Mount Farm [4703 1582]. CRB

The Large-'Paludina' limestone has given rise to a slight north-facing scarp feature between Plashett Park Farm [4560 1457] and near Old Barn Farm [4696 1419]. The limestones were not exposed but loose blocks in the soil, especially in the field [4632 1415] 500 m SSW of Old Barn Farm, are up to 0.15 m thick. A well [4624 1429] on the north side of Green Lane, 530 m west of Old Barn Farm, is said to have been dug through large slabs of limestone up to 0.1 m thick at a depth of about 4.6 m.

About 320 m NW of Broyle Mill Farm a ditch [4684 1385] showed an irregular layer of clay-ironstone nodules 0.05 m thick, underlain by 0.3 m of pale grey clay with slight red mottling and overlain by 0.6 m of fawn and grey clay.

Brick clay was formerly extracted from extensive shallow pits west and south-west of Upper Lodge [4816 1464]. The brickworks closed in 1971, and the only section [4768 1428] visible in 1972, 140 m NW of the works, showed up to 1.2 m of poorly laminated grey and fawn clays. The former works foreman reported that fine- to medium-grained soft brown sandstone in beds up to 0.08 m thick was occasionally encountered in working the pit and some sandy

ironstone fragments were noted in the spoil, though none was seen *in situ*.

The Large-'Paludina' limestone (Bed 4) extends eastwards from a point 1.1 km east of Broyle Mill Farm [4815 1373] capping a relatively well marked south-facing scarp. This scarp becomes increasingly prominent towards Laughton Lodge [4942 1329] and blocks of limestone up to 0.2 m thick with abundant large *Viviparus fluviorum* shells are common in the field [4872 1341] 700 m west of the lodge. An old pit [4918 1335] 220 m west of Laughton Lodge was presumably opened for Large-'Paludina' limestone, which is known locally as 'Laughton Stone'. BY

The deep Laughton Well [4948 1367] proved the sequence shown in Figure 8 and terminated in Ashdown Beds at 326 m. The correlation, which is based on the top of the Wadhurst Clay as datum level, indicates that the adjacent outcrops of marker clay belong to Bed 3. RDL

Area between Laughton and the Cuckmere River

Strong strike-faults affect the Weald Clay between Laughton and Lower Dicker so that where distinctive limestones are absent the marker clays cannot be distinguished. Many small pits were dug for 'marl' in these clays, which at surface are typically greyish green with patchily red upper parts locally. The beds above the marker clays tend to be very silty and to contain thin impersistent sands.

A well near Gildridge [525 140], within the crop of a clay-marker, recorded 'Cyrena' beds below 5.5 m depth, suggesting that this clay belongs to Bed 3. A number of pits were dug in this clay near Kiln Wood [532 130] for brick clay and possibly also to work an ironstone which occurs below it. Medium grey shaly clay with thin siltstones was noted beneath 1.5 m of weathered clay in a cleared face near Kiln Wood [5306 1319].

Bed 4 (Large-'Paludina' limestone) has been worked in pits southeast of Stone Cross [5120 1263], at The Stall [5170 1196], and at Tile Barn [5415 1060]. The walling of parts of Marchants Farm at Stone Cross is partly constructed of this rock (Plate 4), as White recorded (1926, p.29). Mapping of the marker clay associated with this limestone demonstrated the three-fold repetition of the sequence by strike-faulting.

Anderson (*in* Reeves, 1958, p.7) erected the Hackhurst fauni-cycle (lower Weald Clay) on the basis of the fauna obtained from a former brickpit [561 123] near the farm of that name. North of the lowest red marker clay unit which runs parallel to and just south of the road at Lower Dicker, greyish green and pale grey clays show only rare red mottling. White (1926, p.29) described a former brickpit north-east of Lower Dicker village (possibly [558 119]) as working red and mottled clays.

Bed 3 apparently underlies much of the country between the Lower Dicker road and Upper Dicker and brickpits were formerly worked in this area. In the field east of Dicker Pottery [5680 1114], Kirkaldy and Bull (1936) recorded fine-grained sand overlying red sandy clay. Clay with Large-'Paludina' limestone crops out immediately north of The Dicker; White (1926, p.29) recorded an occurrence of this bed 1.2 km north of Upper Dicker but it was not located in the recent survey. RDL

The two pits of the now-abandoned Cuckmere Brickworks north of Berwick Station formerly provided important sections in the Weald Clay and Lower Greensand. The pit 890 m north of the station [5269 0766] is now flooded and the northern face showed only 1 m of pale grey clays and silty clays. At the north-east end of the more extensive old pit [5250 0710] immediately north of the station, up to 1.8 m of grey and fawn clays and silty clays were observed, the pit apparently being floored by a clay-ironstone 0.15 m thick. Kirkaldy (1937) described a section in this pit showing the Weald Clay overlain with slight angular discordance by glauconitic loams with a basal phosphatic pebble bed assigned to the Sandgate Beds. This section has long been obscured.

Plate 4 Large-*'Paludina'* limestone from the Weald Clay used for building a barn wall at Stone Cross, near Laughton (A 12275)

Grey and fawn clays and silty clays with several ironstones up to 0.15 m thick were noted at various points along the north-west banks of the Arlington Reservoir [5312 0766]. Several of these ironstones contained abundant *Filosina sp.* shells and one contained small *Viviparus sp.* BY

In 1969 the borrow-area [532 077] for the construction of the Arlington Reservoir showed the following succession of beds beneath 1.8 m of Lower Greensand:

	Thickness m
Clay, silty, greenish grey, ochreous mottled, weathering white	up to 1.2
Clay, silty, greenish grey, silt-laminated	0.15
Mudstone, greenish grey, ochreous, ?structureless, reddish in top 0.1 m; ostracods	0.5
Mudstone, greenish grey, finely silt-laminated; possibly slipped in part	about 1.2
Mudstone, dark grey, cut by slip planes	0.6
Siltstone, calcareous, fish debris on upper surface	up to 0.1
Mudstone, dark greenish grey; burrowed near the gradational base	0.15 to 0.2
Mudstone, medium grey, finely silt-laminated	about 5.0
Limestone, pyritic, with *Filosina sp.*	0.1

The presence of low-angle slip-planes suggests that this is not the complete stratigraphical succession. Below, pale to medium grey silt-laminated mudstones were noted with two siltstones 0.25 and 0.05 m thick at estimated depths of 7.6 and 10.7 m below the limestone. Beds lower in the sequence were seen in the excavation for the northern part of the dam (Figure 31) and included red-mottled mudstones and associated grey sandy mudstones disturbed by valley-bulging. A uniformly red mudstone was also noted. The best stratigraphical control for this group of beds was provided by a trial borehole [5312 0700] which was sited 55 m from the Lower Greensand outcrop (and topographically 5 m beneath it). This borehole proved a red and grey clay below 20.4 m depth and overlying multi-coloured clay which extended down to 23.2 m. On this evidence, and from a comparison with the Ripe Borehole, it appears that the exposed strata lie considerably above Bed 4. RDL

Area east of the Cuckmere River

The road cutting on the Hailsham bypass [5766 1027 to 5776 1000] displayed the following section in 1971, the beds dipping gently southwards:

	Thickness m
Silts, clayey, ochreous pale grey, with sandy intercalations, passing down to	1.2
Sand, fine-grained, friable, ochreous, with fairly sharp base	0.3
Silts, clayey, ochreous pale grey weathering to buff and reddish tinted; local sandy intercalations	1.5

	Thickness m
Clay-ironstone, locally splitting into two	0.08
Silts, clayey, pale grey, buff locally; bioturbated with relict laminations	2.4
Clay-ironstone	0.05
Clay, silty, ochreous pale grey	2.4

The red clay marking the top of Bed 3 crops out at the road junction to the north [5762 1038]. The Hailsham Borehole [5746 1083] was drilled some 450 m NNW of this point and proved the sequence from Bed 3 to middle Tunbridge Wells Sand (Figure 8; Lake and Young, 1978).

Reeves (1958, p.6) described 'varied grey shales, clays and ironstones' at the Rucklands Brickworks [583 107], with one ostracod-bearing purple shale horizon. The pit was situated in beds below the marker clay at the top of Bed 3. Greyish green clays that are extensive in the Harebeating area [594 107] are grouped with Bed 2.

Hailsham Brickworks [591 087] at one time exposed grey clays dipping at 3°NE. Mr A. H.Taitt (in MS) recorded a 'conglomerate band of very ferruginous clay pebbles set in a ferruginous sandy matrix' in the upper part of the face. He also described the former Coldthorn Brickworks section [585 080] as showing heavy blue clay weathering rust-coloured with some thin irregular ferruginous sandy bands; the dip was approximately 5°SW. Both of these sections showed beds below Bed 4.

No Large-'*Paludina*' limestone has been recorded in the area east of the Cuckmere River, although the Bed 4 marker clay which crops out extensively between Upper Dicker and Polegate is considered to lie at this horizon. The Southdown or Keymer Tileries [583 069], near Coppards, formerly exposed contorted beds at the level of the marker clay; Reeves (1958, p.6) described varicoloured clays (pale grey to chocolate) with ironstones and siltstones. The section recorded south of the disturbed zone in 1971 showed:

	Thickness m
Clay, ochreous, pale grey, with an ironstone bed 0.03 m thick 2.1 m from the top	3.7
Clay, red and pale grey-mottled, weathering red	0.6
Clay, pale grey	0.6
Clay, purple and grey-mottled	0.2
Clay, pale grey, locally colour-mottled	0.9
Clay, red (marker clay)	—

The pit was more recently (1973) worked down to the red clay, which showed a north-eastward dip. Older pits, now degraded, were dug below this level.

Mr Taitt (in MS) described the former Wilmington Brickworks section [5535 0700] as consisting of shaly-weathering blue clay, rust-coloured, with several ironstone bands towards the base; the dip was recorded as 2° to 3°SW. This section lies stratigraphically above the marker clay (Bed 4). Similar beds were also recorded at the former Polegate Brickworks [5770 0515]. In the road cut north of this locality to opposite Cop Hall [5771 0570], Taitt recorded in descending sequence:

Grey-green clays with occasional ironstone nodules, and thin sandy ferruginous beds
Grey sandy clay with an ironstone band
Soft silty sands with greyish green banded clays with ferruginous layers
Red clays with an irregular black sandy ironstone band on top [marker clay].

In the Pevensey area, interpretation of the stratigraphy is complicated by extensive faulting and the lack of distinctive lithological criteria. A red clay marker bed is present from near Dittons [598 049] to Mountney Level [6306 0395].

A well at Stone Cross just north of the outcrop of the marker clay proved 58 m of clay before penetrating 1.5 m of sand. Pevensey No. 2 Borehole [6075 0455], near Hankham Place Lodge, proved about 67 m of dominantly grey clays overlying sands and silts with intercalations of mauve and red mottled clays to 93.6 m depth. The topmost 12 m of strata included mottled clays and this borehole started within the clay marker; tentative correlations with the Ripe and Hailsham boreholes suggest that the marker clay is at the top of Bed 3 and that both of the boreholes in question penetrated Tunbridge Wells Sand. The occurrence of mottled clays within the Tunbridge Wells Sand of the Pevensey No. 2 Borehole is anomalous, however, although abundant mottled clays have been recorded lower in that formation (see p. 20).

Mr Taitt (in MS) recorded blue clays with sandy ferruginous bands dipping at 12°SSW in the brickworks [607 044] near Dittons.

Reeves (1958) described the railway cutting section [620 040] near Friday Street as showing 'alternate red and yellow clays', although only one homogeneous red clay has been proved by surface-augering in the area.

A red clay (equivalent to Bed 4) crops out [around 620 032] 200 m north of the Lower Greensand at Willingdon Drove and this is correlated with a similar bed in the Hampden Park Borehole (Figure 8) 19 m below the base of the Lower Greensand. Large-'*Paludina*' was not found in the borehole cores. A borehole southwest of Rickney [6214 0664] proved 30 m of Weald Clay overlying 36 m of Tunbridge Wells Sand (Taitt in MS).

At the eastern end of the Lewes district the basal Weald Clay is present at Rockhouse Bank. A borehole [6769 0562] on this 'island' within the levels proved 30 m of Weald Clay overlying 37 m of Tunbridge Wells Sand.

RDL

CHAPTER 5

Cretaceous: Lower Greensand

GENERAL ACCOUNT

Marine influences, which affected the Wealden Basin during later Weald Clay times, culminated in a marine transgression that led to the deposition of the shallow-marine sands, silts, and clays of the Lower Greensand. Casey (1961) reviewed the fauna of this group and proposed a scheme of ammonite zones (see Figure 9). In the Weald generally, the Lower Greensand has been subdivided into four formations: the Atherfield Clay locally at the base, overlain in ascending order by the dominantly arenaceous Hythe Beds, Sandgate Beds and Folkestone Beds. White (1926, p.34) noted that these divisions became less distinct as they were traced eastwards through Sussex, and recent work has shown that this subdivision cannot be applied in much of Sussex (Young and others, 1978, p.B95). In the Lewes district it has not been possible to map separate divisions in the Lower Greensand. The 1:50 000 sheet, in common with previous editions, therefore shows undivided Lower Greensand, although locally the following succession may be recognised in descending order:

Homogeneous fine-grained sands
Grey sandy clays and silts
Green glauconitic clays with phosphatic nodule bed at base.

Throughout the Lewes district the Lower Greensand rests sharply on the silty mudstones of the Weald Clay. Palaeontological evidence from the Ripe [5059 1052] and Hampden Park [6120 0204] boreholes, indicates the presence of a slight unconformity at the base of the Lower Greensand. Dr F.W. Anderson (personal communication) estimated that as much as 10 m of ostracod-bearing mudstone present at the top of the Weald Clay at Ripe might have been removed at Hampden Park prior to Lower Greensand deposition. Kirkaldy (1937) described an exposure of this unconformity in the abandoned Cuckmere Brickworks pit [5250 0710] at Berwick Station (see Figure 10), though the section is now obscured. There was evidence here of erosion and winnowing in the form of a condensed Atherfield Clay fauna within the glauconite-rich basal beds of the Lower Greensand (p. 38; Casey, 1961, p.556). The occurrence of two beds of fuller's earth in the Hampden Park Borehole (p.40) indicates that volcanic ash formed secondary bentonites and bentonitic clays within the Lower Greensand deposits (Jeans and others, 1982).

Much of the Lower Greensand outcrop is concealed beneath a mantle of Head; exposures are few and those that do exist are mainly in the top few metres of the formation. The glauconitic clays, silts and sandy clays at the base probably correlate with the Hythe and Sandgate beds elsewhere; a limited macrofauna was obtained from the Ripe Borehole and comparison with the sequence in the Hurlands Farm Borehole [SU 9413 2104] near Selham, West Sussex, suggests that the greater part of the Lower Greensand at Ripe is of late Aptian age (see Figure 9 and p. 37). Throughout

much of the Lewes district a few metres of fine-grained sands or sand rock, the equivalent of the Folkestone Beds of other areas, appear to be present at the top of the Lower Greensand. Above the highest sands there is a bed of phosphatic nodules which in the Lewes district contains a fauna similar to that at the base of the Folkestone Beds at Folkestone (Casey, 1961, p.560). The succeeding glauconitic sandy clays in this district are locally capped by a further nodule horizon. These beds are shown as a condensed 'Gault' sequence in Figure 9 and they have been mapped with the bulk of that formation; they are described in Chapter 6.

Along the southern edge of the Weald the Lower Greensand thins eastwards towards the coast at Eastbourne. At Streat [3492 1485], about 4 km west of the district boundary, the total thickness of the Lower Greensand is 81 m. At Ripe [5059 1052] it has thinned to 50 m and in the Hampden Park Borehole [6120 0204], about 0.5 km south of the Lewes district, only 20.75 m of Lower Greensand are present. The overall thinning of the beds beneath the supposed equivalent of the Folkestone Beds is accompanied by an eastward lateral passage from predominantly fine-grained, slightly glauconitic, silty sandstones, typical of the Brighton district, to sandy clays and silts, commonly glauconite-rich, which characterise the sequences in the Ripe and Hampden Park boreholes. The upper sand and sandstone unit, taken as the equivalent of the Folkestone Beds, is appreciably finer grained in the Lewes district than in the Brighton district; these sands are absent at Hampden Park, though whether this is due to lateral facies change or to erosion and overstep by the base of the succeeding Gault is not clear.

The thinning of the Lower Greensand when traced eastwards to the Hampden Park area may in part reflect the effect of contemporaneous uplift along the Paris Plage or Pevensey ridge structures (Taitt and Kent, 1958; Lake, 1975b; Young and Monkhouse, 1980). Movement along these structural lines prior to the main onset of Lower Greensand sedimentation may account, at least in part, for the sub-Lower Greensand unconformity proved between Ripe and Hampden Park.

The Lower Greensand has very little topographic expression in this district. Most of the outcrop forms an extension of the Weald Clay lowland, but locally the top sand division has produced a low scarp, capped by the basal beds of the Gault.   BY

DETAILS

Area west of the River Ouse

Topographic rises north of the main outcrop of the Lower Greensand near Wickham Barn and Shelley's Folly [3910 1505; 402 151; 408 150] are capped by glauconitic clay and sandy clay, but only the first of these deposits appears to be undisturbed.

Two wells at Cooksbridge [4008 1329; 4008 1350] recorded hard

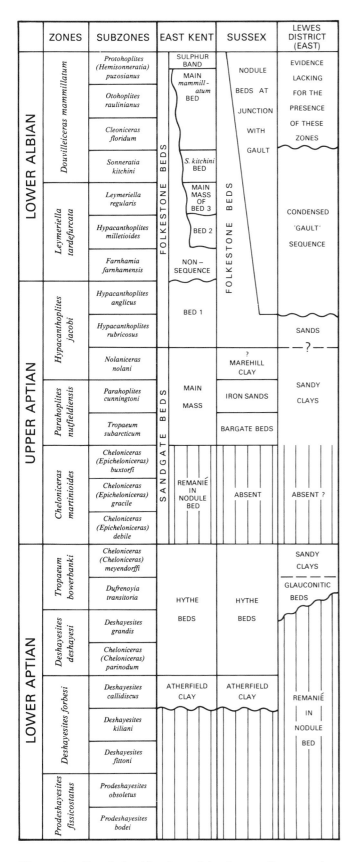

Figure 9 Zonal classification of the Lower Greensand and lowest Gault. Based on Casey (1961)

sandstones within the upper part of the Lower Greensand below the Gault. These indications of cemented beds are unusual.

River Ouse to the Cuckmere River

South of Clay Hill [445 144] the distinctive richly glauconitic clay marks the basal part of the sequence; the bulk of the outcrop is underlain by pale grey sandy clays and clayey sands. The top division of medium-grained sands apparently has an outcrop only about 100 m wide near Ham Farm [4396 1320]; sands were formerly worked at this locality and a small degraded pit [4474 1311] at Norlington exposed buff and red sands showing a tendency to coarsen upward.

The Glyndebourne Borehole [4420 1141] proved the sequence:

	Thickness m	Depth m
HILLWASH AND LOWER CHALK (see p.67)	48.36	48.36
GAULT (see p.41)	104.24	152.60
LOWER GREENSAND		
Sandstone, medium-grained, dark green, glauconitic, friable; quartz pebbles up to 5 mm across; dark purplish brown clay mottles	2.90	155.50
Core loss in soft sand	1.60	157.10
Sandstone, as above but without pebbles; burrow traces; local clay intercalations	4.67	161.77
Core loss in soft sand	5.26	167.03
Siltstone, clayey, medium dark grey, with sandy intercalations; bioturbated; compact, less sandy below 170 m; pyrite nodules	seen to 3.88	170.91

The beds between 152.60 and 167.03 m probably represent the Folkestone Beds.   RDL

The basal glauconitic clays and overlying clayey sands were traced relatively readily eastwards from near Broyle Mill Farm [4704 1358] to south of Laughton [5000 1200]. A small degraded pit [4886 1203] 600 m south of Colbrands was dug through the basal Gault to work the sands at the top of the Lower Greensand.   BY

The outcrop of the topmost sand division extends along the low ridge on which Ripe [510 100] and Chalvington [520 094] are situated, and in the past many small pits have been dug in these beds. An exposure near Manor Farm [5076 1016], Ripe, is described on p.42.

The following sequence was proved in the Ripe Borehole [5059 1052]:

	Thickness m	Depth m
HEAD	1.50	1.50
LOWER GREENSAND		
Sandstone, fine-grained, yellowish brown, friable, increasingly clayey downward	4.16	5.66
Mudstone, sandy, and clayey sandstone, dark grey, tinged purple locally, pyritic, variably glauconitic, bioturbated	20.84	26.50
Mudstone, sandy, calcareous, dark grey, tinged purple locally, variably glauconitic, bioturbated; argillaceous sandy limestone between 40.10 and 40.40 m depth	15.25	41.75
Mudstone, strongly glauconitic, medium grey and green, variably sandy, with septarian nodules and chert nodules, non-calcareous in middle part; pellet bed above burrowed base	6.04	47.79
WEALD CLAY (see p.22)	—	—

The top sandstone may be correlated with the Folkestone Beds of other areas.   RDL

The following bivalves were collected from the beds between 39 and 43 m depth: *Camptonectes sp. ?cottaldina* gr., *Linotrigonia*

(*Oistotrigonia*) cf. *ornata*, *Parmicorbula striatula*, *Pseudolimea* cf. *parallela*. Within the glauconitic beds, the belemnite *Neohibolites ewaldi clava* was noted at 43.45 m, which may indicate the highest Lower Aptian of the *bowerbanki* Zone. *Neohibolites ewaldi* was abundant between 47.00 and 47.53 m. This occurrence invites correlation with the beds below the Coalman ragstone course at Maidstone (Casey, 1961, pp.538–540) and with the beds from 63 to 65 m in the Hurlands Farm Borehole [SU 9413 2104]. A small fragment of *?Nanonavis sp.* collected from the basal bed at Ripe (below 47.53 m) was comparable to several specimens noted in the Hurlands Farm Borehole at 66 to 67 m. In the latter borehole the ammonites indicate that these beds can be ascribed to the *Dufrenoyia transitoria* Subzone of the *bowerbanki* Zone (see Figure 9). The correlation with this and other boreholes in Sussex indicates that the lowest sandy clays of the Lower Greensand in the Ripe sequence are of Lower Aptian age, but most are probably of Upper Aptian age.   AAM

The large old pit [517 095] 300 m NW of St Bartholomew's Church, Chalvington, formerly provided the following section, quoted from Kirkaldy (1935, p.520) with thicknesses converted to metres:

	Thickness m
GAULT (see p. 44)	
LOWER GREENSAND	
'Folkestone Sands'	
White medium sands faintly speckled with green grains of glauconite	[0.46 to 0.91]
Brown sand with scattered quartz pebbles up to 0.5 cm in diameter	[0.61]
White medium sands faintly speckled with green grains of glauconite	[1.83]
Greyish, slightly clayey, sand with a few soft nodules of ironstone	[0.61 seen]

A small section in the southern end of the old, flooded pit [5215 0919] approximately 200 m ESE of Chalvington church showed up to 1.5 m of medium- to coarse-grained, uncemented, slightly micaceous, cream sand with scattered almost black glauconite grains and no sign of bedding. The top 0.6 m of the sand was rather clayey and ferruginous, with abundant quartz grains up to 2 mm in diameter.

The large sand-pit [5250 0900] at Poundfield Corner, 700 m ESE of Chalvington church, is now flooded and overgrown. Kirkaldy (1935, p.521) described just over 2 m of fine- to coarse-grained sand immediately below the Gault.

Two small outliers of the basal glauconitic clays cap the higher ground 150 m west [5462 0947] and 200 m SE [5486 0929] of High Barn.

A trench [5122 0745] 540 m NNE of Selmeston church exposed 0.6 m of very dark grey silty clay with glauconitic sand wisps and burrow fillings, belonging to the middle division, overlain by 1.5 m of pale brown sandy Head. Other excavations in the road at Church Farm, Selmeston [5110 0719], revealed ochreous mottled grey, micaceous, fine-grained sand with clay wisps, probably belonging to the highest part of the middle division.

The top sand division has been worked in a number of old pits east of Selmeston church. Up to 1.5 m of pale fawn, friable, slightly micaceous, fine-grained sand was noted in the north side of an old pit [5123 0691] about 250 m SE of the church; the sand was formerly dug to a depth of 3.7 m (Plate 5; White, 1926, p.36). Casey (1950, p.279) described the section exposed in this pit as follows: 'The sands are predominantly pale yellow, buff, and grey in colour, evenly bedded and of fine texture. Buff sand shot with veins of dark green sandy clay like that at the bottom of the Gault was visible near the floor of the pit in 1949, and rafts of mauve and grey clay were seen in the three feet of sand above. For the next ten feet the sand

carries argillaceous matter only in wisps. This part of the section is the source of the friable ironstone nodules lying about the pit. Glauconite is disseminated throughout the sands and is particularly abundant just below the base of the Gault.'

'This section does not show the contact with the Gault in place, but in the south-east corner of the pit the sands are overlain by a bed of phosphatic nodules — a residuum from the erosion of the feather-edge of the Gault. Were it not for the presence occasionally of a flint or some other foreign body among the nodules, the bed could easily be mistaken for the base of the Gault *in situ*. *Hypacanthoplites* and other fossils were obtained from the nodules.' Casey (1950, p.279) also described 1.8 m of sands beneath the Gault, no longer exposed, in the overgrown pit [5108 0689] 120 m SE of Selmeston church.

A well at the Reading Room [5083 0681], Selmeston, passed through the base of the Gault into the top sand, proving at least 10.67 m of sand.

The extensive old pit of the abandoned Cuckmere Brickworks [5250 0710], immediately north of Berwick Station, formerly provided important sections in the Weald Clay and basal beds of the Lower Greensand. Kirkaldy (1937, p.106) gave the following description of the section available in 1934, including evidence of discordance of the Weald Clay–Lower Greensand junction, with the basal Lower Greensand nodule bed over-stepping ironstone beds within the Weald Clay (Figure 10):

	Thickness m
HEAD	
Brown sandy brickearth, glauconitic in its lower part, containing seams and pockets of angular flints	[0.61 to 0.91]
LOWER GREENSAND	
Glauconitic greenish and reddish coloured loams, the well-marked bedding being emphasised by the browner bands. Scattered phosphatic nodules up to 0.08 m in diameter occur in its lower part	[1.83]
Dark olive-green coloured highly glauconitic loams with yellow patches	[2.74]
Seam of small greyish and greenish-coated phosphatic nodules up to 0.05 m in diameter	[0.08 to 0.15]
WEALD CLAY	—

This basal remanié nodule bed of the Lower Greensand, which incorporates not only Weald Clay material but also derived Lower Aptian ammonites including *Prodeshayesites sp.* and *Deshayesites forbesi*, clearly represents a considerable non-sequence. Casey (1961, p.556) suggested that the bed was a highly condensed

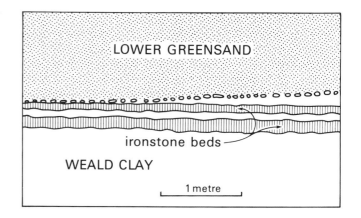

Figure 10   Sketch of a section in Weald Clay and Lower Greensand in Cuckmere Brickworks pit. After Kirkaldy (1937)

Plate 5   Folkestone Beds sands in a pit south-east of Selmeston church (A 2979)

remanié of the whole of the Atherfield Clay. White (1926) noted the occurrence of lignite fragments within dark olive-green loamy sand above the nodule bed. The pit is now overgrown and the only section still exposing Lower Greensand at the time of survey showed 1.5 m of medium grey clay with silt laminae (Weald Clay), overlain by 0.9 m of dark green glauconitic clayey silt.

A small pit [5342 0614] south of the railway on the west bank of the Cuckmere River provided the following section in the top sandy division:

	Thickness m
HEAD	
Loam, sandy, brown, with pockets of flint gravel	up to 0.9
LOWER GREENSAND	
Sand, fine- to medium-grained, cream, bioturbated with clay wisps	0.6
Clay, dark grey, with burrows filled with sand similar to that above	0.4
Sand, fine- to medium-grained, cream, bioturbated with clay wisps	up to 1.2

Area east of the Cuckmere River

Another small and much degraded pit [5389 0595] immediately south of the railway on the east bank of the Cuckmere River exposed up to 1.5 m of the top sands overlain by a bed of phosphatic

nodules, 0.05 m thick, possibly the basal nodule bed of the Gault.

Dark green glauconitic silty clays were seen in the banks of the lane [5427 0658] 240 m NE of Endlewick Farm.   BY

East of Wilmington, the basal glauconitic clay beds, occurring on a dip-slope, increase their outcrop area at the expense of the middle sandy clay division, whereas the top sands seem to be absent north-east of New Barn [556 052]. East of Wootton Manor [565 052] the outcrop of the Lower Greensand narrows still further, although the top sands are present. Kirkaldy (1935, p.523) recorded the basal nodule bed of the Gault just east of the railway bridge [5689 0481] and Topley (1875, p.127) described the section in the adjacent railway cutting as:

	Thickness m
GAULT	
Green rather clayey sand	[0.46]
Phosphatic nodules and wood	[0.10]
LOWER GREENSAND	
Red sand	[0.30]
Sand, getting greenish below	—

Between Polegate and the coast, the basal glauconitic clay has been recorded in temporary exposures opposite the Horse and Groom Inn [5805 0459] and in Brightling Road culvert cut [5833 0442]. The road cutting [5826 0422] north-east of the windmill has on occasions exposed 'Folkestone Sands' overlain by olive-green glauconitic sandy clay with the basal nodule-bed of the Gault clay, dipping eastwards at 5° (Kirkaldy, 1935, p.523; Casey, 1961, p.560). A sewer excavation [5950 0387] north-west of Brodrickland Farm proved the basal contact with the Weald Clay beneath 1.5 m of loamy Head. Up to 2.4 m of the basal glauconitic sandy clays were recorded in a temporary excavation [6199 0288] in Willingdon Drove, east of the Willingdon Levels.   RDL

In the Hampden Park Borehole [6120 0204], approximately 0.5 km south of the boundary of the Lewes district, an attenuated Lower Greensand sequence only 20.75 m thick was proved. The succession is summarised as follows:

	Thickness m	Depth m
ALLUVIUM	6.00	6.00
GAULT (see p. 45)	12.82	18.82
LOWER GREENSAND		
Sand, fine-grained, clayey, olive-grey, slightly micaceous, a little glauconite, strongly bioturbated, some minor siltstone beds	6.33	25.15
Siltstone, sandy, clayey, olive-grey, glauconitic, strongly bioturbated	2.45	27.60

	Thickness m	Depth m
Fuller's earth, olive-grey, massive; burrowed top, sharp base	0.23	27.83
Sand and sandstone, very fine-grained, olive-grey, glauconitic, locally calcareous, bioturbated, a few minor siltstone beds	2.44	30.27
Fuller's earth, olive-grey, massive; sandy top, sharp base	0.32	30.59
Sand and sandstone (as above)	0.98	31.57
Siltstone, clayey, locally sandy, greenish grey, glauconitic, calcareous, bioturbated, phosphatic nodules locally in lower part; sharp base	8.00	39.57
WEALD CLAY (see Figure 8)	50.53	90.10

The top 6.33 m of sand do not resemble the top sand division correlated with the Folkestone Beds elsewhere in the Lewes district and are more like the beds beneath this division in the Ripe Borehole. The Hampden Park Borehole therefore appears to support Casey's suggestion (1950; 1961) that the 'Folkestone Beds' are absent in the Eastbourne area.   BY

CHAPTER 6

Cretaceous: Gault

GENERAL ACCOUNT

The Gault, which occupies a low-lying belt of country at the foot of the Chalk escarpment, consists mainly of medium grey, pyritic, silty mudstones which weather at the surface to bluish grey or pale grey clays. At the base distinctive phosphatic nodules and lignite occur within glauconitic sandy clays which constitute the 'basement-bed', directly overlying the sands of the Lower Greensand (see Figure 9 and p. 36). This boundary has been shown to be diachronous (Casey, 1961) and marks a transgression of the Albian sea.

Layers of phosphatic nodules also occur higher in the succession, representing pauses in deposition. Thin beds of argillaceous limestone and of shelly (bivalve) limestone are also present.

Near the top of the formation the mudstones pass upward into very silty, calcareous, micaceous beds; the more silty beds have been previously called 'malm', a term which was applied to the unconsolidated equivalent of the 'malmstones' described in the Upper Greensand elsewhere in the Weald. At outcrop in this district the upper beds of the Gault do not become sufficiently silty to justify separation. However, on the coast to the south at Eastbourne, some 10 m of medium-grained sandstone occur at this level, indicating that concealed deposits of Upper Greensand are present to the south of the Lewes district.

A slight unconformity at the top of the Gault, beneath the Glauconitic Marl of the Lower Chalk, has been observed in a clay-pit at Rodmell [441 071], and supporting evidence has been provided by the microfaunal zonation (see below).

The Glyndebourne Borehole [4420 1141] proved the full Gault sequence which is summarised as follows (see also Appendix 3):

	Thickness m	Depth m
LOWER CHALK: Glauconitic Marl (see p.66)	—	48.36
GAULT		
Mudstones, very silty, medium grey, bioturbated, calcareous, with subordinate clayey siltstone	18.64	67.00
Mudstones, silty, medium dark grey, less calcareous than above; shell debris common, muddy limestones up to 0.11 m thick	48.30	115.30
Mudstones, shelly, medium dark grey; glauconitic intercalation 1.67 m from top	33.16	148.46
Mudstones, glauconitic, quartzose at base	4.14	152.60
LOWER GREENSAND (see p. 37)	—	—

The thickness of 104 m is the greatest known in the Weald for the Gault.

In the Glyndebourne Borehole two clay mineral assemblages have been identified in the Gault (Jeans and others, 1982). A mica-kaolinite assemblage characterises clays below about 84 m depth, whereas a smectite-mica assemblage is present higher in the sequence.

The Gault contains a varied marine fauna of bivalves and ammonites, particularly in the lower part. Gastropods, brachiopods, foraminifera, fish fragments and drift wood also occur. The division of the Albian into ammonite zones (following Owen, 1976; 1979; 1984) is illustrated in Figures 9 and 11 and tabulated below:

Stage	Zone	Subzone	Thickness[1] m
LATE ALBIAN (Upper Gault)	Stoliczkaia dispar	Mortoniceras (Durnovarites) perinflatum	—
		Mortoniceras (Mortoniceras) rostratum	9.49 +
	Mortoniceras (Mortoniceras) inflatum	Callihoplites auritus	29.70
		Hysteroceras varicosum	15.78
		Hysteroceras orbignyi	5.41
		Dipoloceras cristatum	7.31 +
	Euhoplites lautus	Anahoplites daviesi	—
		Euhoplites nitidus	?1.79
MIDDLE ALBIAN (Lower Gault)		Euhoplites meandrinus	2.79
	Euhoplites loricatus	Mojsisovicsia subdelaruei	0.20
		Dimorphoplites niobe	1.21
		Anahoplites intermedius	14.70
	Hoplites (Hoplites) dentatus	Hoplites (Hoplites) spathi	1.41
		Lyelliceras lyelli	0.35 +
LOWER ALBIAN (part) (Lower Gault– Lower Greensand)	Douvilleiceras mammillatum (part)	Pseudosonneratia eodentata[2]	not recognised
		Protohoplites puzosianus[3]	
		Otohoplites raulinianus	

1 Thicknesses as recognised in the Glyndebourne Borehole
2 Included with the Lower Albian (Owen, 1984)
3 There is possibly a non-sequence in England above this subzone

It has been noted previously (p.36) that nodule beds within the basal clays contain a condensed fauna of Upper Aptian to Lower Albian age.

In the details that follow (and see Figure 11 and Appendix 3) comparison is made with the standard section of the Gault at Copt Point [TR 243 365], Folkestone, which was divided into thirteen beds (I to XIII in upward order) by Jukes-Browne and Hill (1900), based on an earlier subdivision by F. G. H. Price (1874) (see Smart and others, 1966, pp. 56–58). A microfaunal zonation based on benthonic foraminifera has been established by R. J. Price (1977) and Carter and Hart (1977), who worked on samples from boreholes in the Dover area. The latter authors subdivided the Middle and Upper Albian into zones 3, 4, 4a, 5, 5a, 6, 6a (enumerated upwards) and concluded that the Albian–Cenomanian boundary is almost transitional in some parts of south-east England but lies at a marked hiatus in others. In this district they noted a palaeontological break at Rodmell (Asham) [441 071] and West Firle (see below), where the highest beds of the Albian contain a Zone 6 fauna (see also p.49) and their Zone 6a is missing, although the latter is present in the Alciston Borehole (see also p.44) (Dr D. J. Carter, personal communication) and probably represents the higher part of the *perinflatum* Subzone.

In the Rodmell No. 1 Borehole, Price (1977) recorded his foraminiferal Zone 8, equivalent to the higher part of Zone 6 of Carter and Hart (1977). In terms of ammonite zonation this probably means that the lower part of the *perinflatum* Subzone is represented at the top of the Upper Gault in most of the sections examined, with the exception of the Alciston Borehole where the higher part of the *perinflatum* Subzone is preserved.

Harris (1982) has erected a complementary zonal scheme based on both ostracods and foraminifera; his zones for the Albian are Ai, Bi, Ci-ii, Di-iii, Ei-ii, Fi-iii in ascending order, and some of his observations on the Glyndebourne sequence are included in Appendix 3.   RDL,AAM,RNM

DETAILS

Area west of the River Ouse

In the railway cutting [404 129] south-east of Cooksbridge, Topley (1875, p.158) noted that 'the beds not far below the chalk are soft, sandy, and rather brownish, with dark markings. They somewhat resemble malm rock which has not become indurated. The lowest beds, seen at the north end of the cutting, are brownish sandy clays becoming darker and more clayey below, the lowest beds exposed being blue and slate-coloured clay, still rather sandy.'

At Kingston near Lewes the Gault crops out in the core of the Kingston–Beddingham Anticline, and around The Brooks this has given rise to a roughly circular outcrop of Gault. Buff and pale grey silty clays, locally containing race, are present below the Glauconitic Marl near Swanborough Nurseries [403 084] and on each of the Rises [416 088; 425 079].   RDL

River Ouse to the Cuckmere River

At the eastern end of the core of the Kingston–Beddingham Anticline, an extensive outcrop of Upper Gault is present. The large clay-pit [441 071] that served the former Rodmell (Asham) Cement Works exposed:

	Thickness m
LOWER CHALK (see p. 66)	1.2
GAULT	
Mudstones, very silty, buff, bioturbated, passing down to	1.8
Mudstones, silty, medium grey, bioturbated	5.0
Mudstones, generally obscured by talus	5.0

A slight unconformity was apparent beneath the Lower Chalk.

The Gault yielded a microfauna of Zone 6 of Carter and Hart (1977, p.82), with abundant *Globigerinelloides bentonensis* which is equivalent to the lower part of the *dispar* Zone, *rostratum* Subzone. However Price (1977, p.87) recorded his Zone 8 from the Rodmell No. 1 Borehole [440 072] situated within the clay-pit; this zone has been equated with the lower part of the overlying *perinflatum* Subzone.

North of Mount Caburn, in the main Gault outcrop, a road-bank [4295 1241] near Ryngmer Park exposed ochreous pale grey silty clays from the topmost beds.   RDL,RNM

The Glyndebourne Borehole [4420 1141] proved the greatest thickness of the Gault known in the Weald and is described in Appendix 3. The zones used in that description of lithologies and palaeontology are based upon the scheme proposed by Owen (1976; 1979; 1984). The subzonal thicknesses are summarised above (p.41). In parts of the sequence (Figure 11) there is insufficient evidence to place firm subzonal boundaries and correlation with the standard sequence established at Folkestone is not possible at all levels.   AAM

South of Ringmer, White (1926) recorded spoil from a trench between Middleham [4424 1195] and Goat Farm [4482 1180]. He took the material to be from the Upper Greensand but it is now perhaps best interpreted as Head, which forms a blanket over much of this area.

The sides of the large pit [4690 0980] 1.5 km NE of Glynde Station are now almost totally degraded. Only about 1 m of grey clay was noted above the talus at the eastern end of the pit. White (1926, p.41) recorded the following Gault ammonites from this pit: *Hamites* cf. *turgidus*, *Hoplites* (*Epihoplites*) aff. *deluci*, *Hoplites* (*Epihoplites*) *denarius*, *Hysteroceras binum*, *H. symmetricum* and *Hysteroceras sp. nov.* Following the re-definition of the *varicosum* Subzone (Owen, 1976), this assemblage is indicative of that subzone, i.e. Bed X and the lower part of the Bed XI of the Gault at Folkestone. The following fauna has also been recorded from this pit by Johnson (1901): '*Nucula pectinata*, *Ancyloceras spinigerum*, *Scaphites hugardianus*, '*Schloenbachia*' *varicosa*, '*Desmoceras*' *beudanti*, tooth of *Protosphyraena ferox*'.

The West Firle Borehole [4676 0809], drilled in 1973, 180 m west of Gibraltar Farm, passed through the Glauconitic Marl into pale grey silty calcareous mudstone typical of the high beds of the Gault. From a study of the microfauna Dr D. J. Carter (personal communication) reported the presence of a palaeontological break at the base of the Chalk where his Zone 6a is absent, and the Chalk rests disconformably on Zone 6. Similar mudstones were seen in excavations [4707 0811] during the construction of the road 100 m ENE of Gibraltar Farm. Up to 2.4 m of grey clays and silty clays were exposed in the excavation for the road cutting between 250 and 900 m east of Wick Street [4765 0808 to 4828 0803].   BY

A temporary exposure [5076 1016] near Manor Farm, Ripe, showed the following sequence across the Lower Greensand–Gault junction:

	Thickness m
GAULT	
Silt, ochreous pale grey, tinged green; gradational base (possibly Head in part)	1.2
Silt, clayey, green, glauconitic	1.2
Sandstone, clayey, reddish brown, with phosphatic nodules and wood fragments at base	0.3

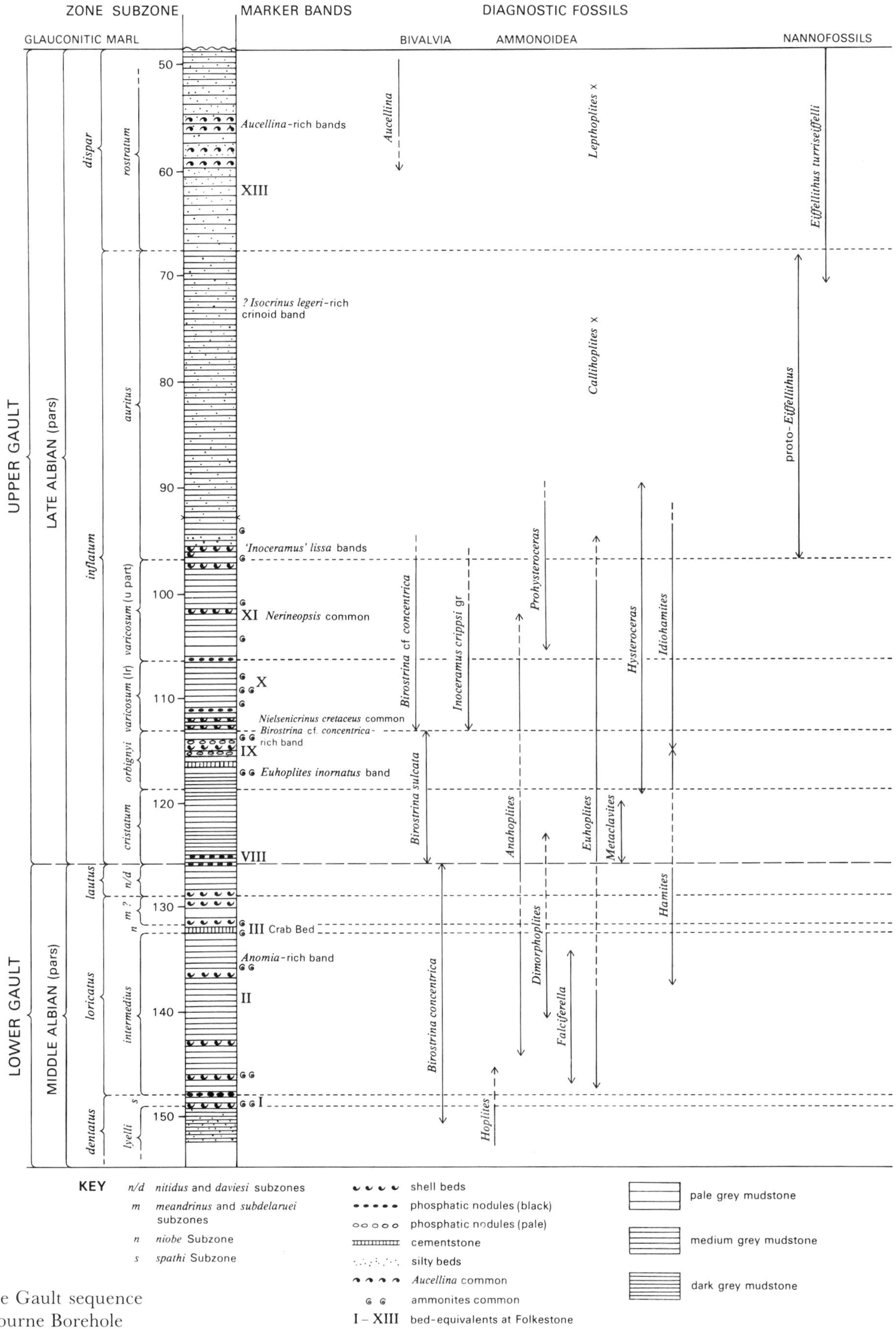

Figure 11 The Gault sequence in the Glyndebourne Borehole

Thickness
m

LOWER GREENSAND
Sand, fine-grained, ochreous pale grey, locally blocky,
consolidated, with vertical iron-staining   3.7

RDL

The spoil from a 5.5-m deep trench [5095 0974] at the sewage works west-south-west of Ripe church showed dark grey clay and some sandy glauconitic clay from near the base of the Gault.

A large sand-pit [517 095] at Manor Farm, 300 m NW of Chalvington church, formerly exposed the junction of the Lower Greensand with the Gault as follows (Kirkaldy, 1935, p.520):

Thickness
m

GAULT
Clay, sandy, green, glauconitic, with lenticles of
reddish brown clay   [0.30 to 0.46]
Discontinuous layer of white phosphatic nodules
and pieces of bored wood   [0 to 0.08]
Clay, sandy, green, glauconitic   [0.46 to 0.91]
Continuous layer of nodules and wood   [0.08 to 0.15]
'FOLKESTONE SANDS' (see p. 38)   —

Casey (1950) recorded the following fauna from the two nodule beds:

Lower nodule bed
Ammonoidea: *Hypacanthoplites jacobi*, *H. clavatus*, *H.* aff. *elegans*, *H.* cf. *spathi*, *H. spp.*
Bivalvia: '*Exogyra obliquata*' (= *E. conica*), *Gervillella sp.*, *Chlamys robinaldina*, *Entolium orbiculare*, *Venilicardia sp.*, *Martesia prisca* in wood, cf. '*Turnus*' *amphisbaena*.
Brachiopoda: '*Rhynchonella*' cf. *sulcata*, *Lamellaerhynchia caseyi*. Also fish scales.

Upper nodule bed
Ammonoidea: *Leymeriella tardefurcata*, *L. regularis*, *L.* aff. *regularis*, *Sonneratia* cf. *sarasini*, *S.* aff. *parenti*, *Cleoniceras* aff. *baylei*, *Douvilleiceras* aff. *mammillatum*, *Douvilleiceras sp.*, *Beudanticeras ligatum*.
Bivalvia: '*Neithea (Neitheops) quinquecostata*', *Martesia prisca* in wood, *Trigonia sp.*

The ammonite fauna of the lower nodule bed provided the first authentic English record of the *jacobi* fauna and showed that the base of the Gault here contains a fauna derived from the *anglicus* Subzone of the *jacobi* Zone (Lower Albian), which occurs near the base of the Folkestone Beds at East Cliff, Folkestone (Casey, 1961). This appears to be a remanié bed, formed by winnowing and sorting of nodules during a long period when no sedimentation took place. The missing interval is considered to be at the base of the Lower Albian *tardefurcata* Zone (Casey, 1950; 1961). The upper nodule bed fauna indicates a remanié of the *regularis* and *kitchini* subzones, comparable with Band II of Leighton Buzzard (Casey, 1961). Clearly a considerable period is represented by the glauconitic sandy clay between the two nodule beds, which Casey (1961, p.529) considered to be of *milletioides* Subzone age and therefore equivalent to the greensand bed (bed 2 of Price) overlying the *anglicus* Subzone nodules at Folkestone.

Kirkaldy (1935) described the following section in the large sand-pit [5250 0900] at Poundfield Corner, 700 m ESE of Chalvington church; the pit is now flooded and overgrown:

Thickness
m

Loamy soil   —
GAULT
Sand, clayey, green, glauconitic, with a thin layer
of nodules at its base   [0.30]
Sand, coarse, glauconitic, current-bedded; the
foresets dip at 27°SE and are picked out by
glauconite grains and white algal markings   [0.46]

Thickness
m

Sand, clayey, well-bedded, greenish brown, with
a considerable amount of glauconite   [0.46]
Thin layer of small nodules   [0.03 to 0.05]
'FOLKESTONE SANDS' (see p. 38)   —

Over much of the pit, Kirkaldy described the upper nodule bed as resting directly on 'Folkestone Sands', but in this measured section the lower beds were a minor channel-fill deposit.

Fossils collected from this pit by Kirkaldy (1935) and by Casey (1950) include: *Douvilleiceras mammillatum*, *Beudanticeras ligatum*, *Sonneratia kitchini*, *Sonneratia sp. nov.*, and *Hypacanthoplites* cf. *elegans*. The two nodule beds here are almost certainly equivalent to the two nodule beds at Manor Farm, Chalvington, and presumably on this basis Casey assigned the species of *Hypacanthoplites* to the lower layer of nodules and the species of *Douvilleiceras*, *Sonneratia* and *Beudanticeras* to the upper nodule bed.

H.W. Bristow (*in* Topley, 1875, p.127) described a sand-pit at Selmeston showing the basal Gault resting on sands at the top of the Lower Greensand with what he described as a 'passage-bed', 0.15 m thick, consisting of grey and brown sands with phosphatic nodules and numerous wood fragments between the Lower Greensand and Gault. This is assumed to be the old pit [5100 0700] immediately north of the churchyard and is believed to be the section from which Mantell (1846) recorded the pine cone *Pinostrobus* [*Zamia*] *sussexiensis*, probably from the nodule bed at the base of the Gault.

Casey (1950) described the following succession in the now overgrown pit [5108 0689] 120 m SE of Selmeston church:

Thickness
m

GAULT
Clay, sandy, olive-green, weathering brown   [0.30]
Band of phosphatic nodules   [0.08 to 0.15]
Sand, clayey, glauconitic, weathering reddish
brown   [0.15]
LOWER GREENSAND
'Folkestone Beds'   —

The nodule bed here yielded *Hypacanthoplites* as well as various bivalves and is equivalent to the lower nodule bed in the Chalvington pits. The basal nodule bed, in a weathered and disturbed form, was also formerly exposed in the south-east corner of an old sand-pit [5123 0691] about 250 m SE of the church (Casey, 1950, p.279).

A BGS borehole at Alciston [5045 0553], in 1973, proved typical pale grey silty calcareous Gault clay immediately beneath the Chalk. Dr D. J. Carter (personal communication) reported a small palaeontological break at the base of the Glauconitic Marl in this borehole: the topmost Gault contains *Aucellina* close to the neotype of *A. gryphaeoides* (see Morter and Wood, 1983) which probably occurs in the lower half of Carter and Hart's (1977) Zone 6a, probable equivalent of part of the *perinflatum* Subzone.

It is likely that some of the old sand-pits [5154 0656; 5220 0636] at the top of the Lower Greensand between Selmeston and Berwick Station exposed the basal beds of the Gault though there are no exposures today. Spath (1930) based his original Sussex record of *Leymeriella regularis*, now *L. (Neoleymeriella) pseudoregularis* on a specimen (GSM 38053) from Berwick Common pits formerly in the collection of Daniel Sharpe (Casey, 1950; 1978). Stopes's (1915) record of the coniferous wood *Protopiceoxylon edwardsi* probably also refers to one of these pits.   BY

Area east of the Cuckmere River

A weathered bed of phosphatic nodules, 0.05 m thick, probably the basal bed of the Gault, was seen overlying 1.5 m of pale fawn medium-grained sand at the top of the Lower Greensand in the small pit [5389 0595] south of the railway on the east bank of the Cuckmere.

On the north side of the A27 road from Polegate to Lewes, between 100 and 300 m east of Sherman Bridge, up to 1.5 m of medium grey clay overlain by flinty Head is exposed in the bank [5340 0504].   BY

In the Hampden Park Borehole [6120 0204] the Gault was penetrated between the depths of 6.00 and 18.82 m, all belonging to the *dentatus* Zone. The beds to a depth of about 14.3 m represent the *spathi* Subzone. The lowest zonally diagnostic fossil recorded was a specimen of *Protanisoceras barrense* at 14.49 m, indicating the *lyelli* Subzone. The zonal position of the lowest 4.33 m of Gault is thus unknown.   AAM

CHAPTER 7

Cretaceous: Chalk

INTRODUCTION

The Chalk crops out in the south-western part of the district and comprises a sequence of predominantly soft, white, very fine-grained extremely pure limestones up to 360 m thick. These rocks consist mainly of coccolith biomicrites formed from the skeletal elements of minute planktonic green algae, associated with varying proportions of larger microscopic fragments of bivalves, mainly *Inoceramus*, echinoderms, foraminifera and ostracods. Traces of clay-grade detritus (mainly quartz) occur throughout the Chalk, but overall comprise less than one per cent of the rock. At various levels clay material occurs as discrete marl seams or diffuse plexi. Locally these may rest on minor erosion surfaces. Many of the marl seams provide persistent marker horizons which can be recognised both at outcrop and, using geophysical logs, in boreholes.

Flint is present as nodular seams, tabular beds and linings to fractures, and characterises the upper parts of the sequence. It comprises a random mosaic of quartz crystals, only a few microns in diameter, interspersed with minute water-filled cavities. The silica was derived from the dissolution of the siliceous skeletons of sponges and other organisms and has been redistributed to form nodules during several stages of crystallisation. The earliest precipitations occurred near organic remains such as burrow-fills or other decaying debris. Further accretion gave rise to either isolated nodules or layers of tabular flint. The preservation of uncompressed fossils in flint indicates that the replacement commenced early in the lithification of the chalk. However, the presence of flint sheets along joints and faults shows that the quartz remained mobile and recrystallised during subsequent earth-movements. The courses of nodular flints closely parallel the bedding and some are valuable for the purposes of correlation.

Phosphate minerals are widespread in the Chalk and localised concentrations occur as impregnations and coatings on the sediments, or as pelletal concretions. Glauconite occurs in minute quantities throughout the Chalk, but is concentrated at some horizons as detrital grains or as encrustations or replacements. Finely disseminated pyrite is common, particularly in the marls, and some of the pyrite forms spheroidal concentrations of radiating crystals.

In areas of relatively shallow-water deposition the upper surface of the sediment may have been secondarily cemented by aragonite, the degree of secondary lithification ranging from patchy hardening to complete induration. Nodular chalks are the first stage in this process of progressive syn-sedimentary cementation, leading to the formation of a homogeneous lithified sediment known as a chalkstone (Bromley and Gale, 1982), capped by a mineralised hardground surface (Bromley, 1978), impregnated by goethite, glauconite and phosphate.

A distinction can be made between lithologies which in the past would have been grouped together under the general description 'nodular chalks' and in particular between griotte chalks and nodular chalks, in a restricted sense (Mortimore, 1979). Griotte chalks have a structure comprising 'augen' of chalk enveloped by marl. This fabric, which has resulted from processes of bioturbation, has been extensively modified by early diagenesis and compaction. Two types of griotte can be recognised: nodular griottes, where the augen are markedly flattened, have angular forms and are crossed by stylolites; and soft-sediment or 'smooth' griottes, where the augen have subrounded surfaces and do not project in weathered surfaces. The term 'griotte marl' has been applied to the more marly chalks which possess this structure. Nodular chalks, in the restricted sense, may superficially resemble nodular griottes but marl envelopes are lacking in the fabric. Phosphatised and glauconitised hardgrounds are found in the Lewes district but generally they are rare.

In parts of the Chalk succession the bedding is obscured by the presence of flasers of clay or marl which impart a wispy appearance to the rock. The flasers occur both parallel to the bedding and at low angles to it. These features are thought to be due to preferential solution which caused segregation of the clay and carbonate in the sediment. Where clay material is absent the effects of syndepositional and post-depositional solution on the Chalk are difficult to assess.

The macrofauna of the Chalk includes planktonic forms such as belemnites, crinoids, foraminifera and some ammonites. These fossils are associated with a bottom-dwelling fauna, including benthonic foraminifera, brachiopods, bivalves and echinoids, notably *Micraster* and *Echinocorys*. During diagenesis, shells with aragonite such as ammonites and gastropods tended to be dissolved, and such fossils are generally only preserved in the lower part of the Chalk. Thus the Chalk fauna is not fully representative and only the calcitic component is commonly preserved.   RDL,RNM,CJW

LITHOSTRATIGRAPHICAL SUBDIVISION

The Chalk has traditionally been divided into Lower, Middle and Upper units (about 86, 95 and 180 m thick respectively in this district). The top of the Lower Chalk is taken at the base of the nodular Melbourn Rock. The subdivision of the Lower Chalk into the Glauconitic Marl, Chalk Marl, Grey Chalk and Plenus Marls (Figure 12) is imprecise and currently subject to review. This sequence comprises mainly rhythmically bedded marls and marly chalks, the calcimetry of which increases upwards in an irregular fashion (see, for example, Destombes and Shephard-Thorn, 1971, fig. 3): e.g. the limestone components of the rhythms are thicker in the Grey Chalk (75 to 85 per cent $CaCO_3$) than in the Chalk Marl (less than 60 to 75 per cent $CaCO_3$). Thus, although these subdivisions are distinctive in gross

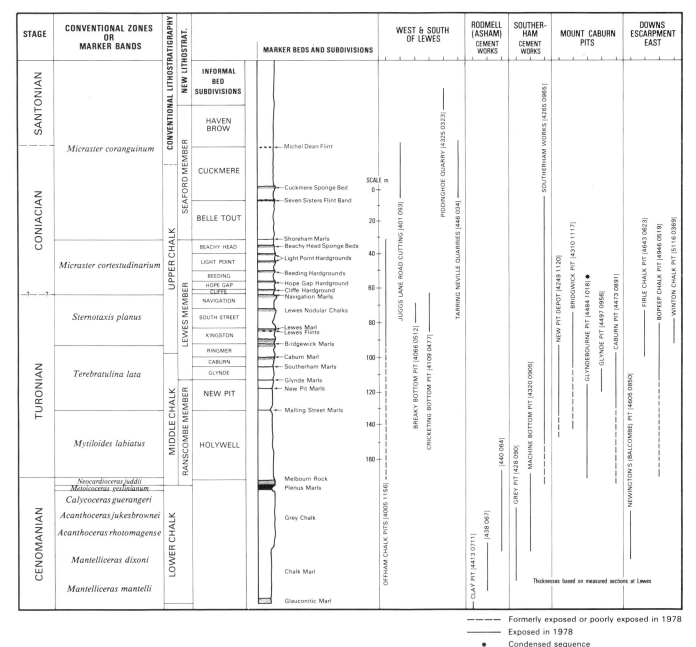

Figure 12   The stratigraphical position of important sections in the Chalk around Lewes

lithology it is considered impracticable to define the intervening boundary precisely in purely lithological terms, particularly in weathered exposures. Faunal marker bands near the gradational boundary are more useful for purposes of correlation (see below).

The boundary between the Middle Chalk and Upper Chalk is taken in the Chilterns and adjacent areas at the base of a sequence of mineralised hardgrounds known as the Chalk Rock, which represents a condensed succession. However this criterion is not applicable in the Weald, where the sequence is less condensed. In the latter area the base of the Upper Chalk has been drawn either biostratigraphically at the base of an ill-defined and subjectively assessed *Sternotaxis planus* macrofossil zone (see below), or lithostratigraphically

using the succession at Dover (Jukes-Browne and Hill, 1904, fig.45) as a standard; there the base of the Upper Chalk was defined at the base of a 1.5-m belt of large nodular flints underlying a conspicuous and closely-spaced pair of marl seams with large nodular flints in between and immediately above (Figure 20). This very distinctive sequence, known as the 'Basal Complex' (Stokes, 1975), is the local expression of a flint maximum which can be traced throughout the English Chalk at this level, and which includes the famous 'Brandon Flint Series' (see discussion by Mortimore and Wood, 1985). At Dover the Basal Complex marks the appearance of flint in quantity into the succession, and approximates to the boundary between the 'Chalk with few flints' below and the 'Chalk with numerous flints' above (Phillips, 1819). Subsequent work by the Geological Survey demonstrated that the belt of large flints at the base of the Basal Complex could be traced throughout the North Downs, and it was consequently possible to define a consistent base to the Upper Chalk over a large area of southern England. However, the dramatic contrast seen at Dover between the Basal Complex flint maximum and the underlying (poorly) flinty chalk does not obtain in the South Downs. Here, belts of large nodular flints are present in places (notably in the Southerham Works Quarry near Lewes) far beneath the local equivalent of the Basal Complex, which is itself not so dominant within the succession as it is at Dover. Because of this, it is impossible to use flint as a criterion in defining and mapping the base of the Upper Chalk in the present area. Attempts to define this boundary on palaeontological criteria have proved equally unsatisfactory; indeed, it has been drawn by Chalk stratigraphers at five different horizons in different parts of the south coast. This inconsistency applies in particular to the previous editions of the memoirs relating to the Lewes district (White, 1926) and to the adjacent Brighton and Worthing district (White, 1924). In the absence of a mappable marker bed at this level the Middle Chalk and Upper Chalk have been shown undivided on the published 1:50 000 geological sheet. However, a detailed investigation of the Middle and Upper Chalk of Sussex, with particular emphasis on the relationship between lithostratigraphy and physical characteristics in a civil engineering context,

coupled with a re-appraisal of the biostratigraphy has recently been carried out by Mortimore (1979; 1983; 1986a). He has divided his 'Sussex White Chalk Formation' (i.e. the 'Upper and Middle Chalk undivided' of this memoir) into six members, of which the two highest are not available for study in the Lewes district:

Portsdown Griotte Marl Chalk Member
Culver Soft Chalk Member
Newhaven Griotte Marl Chalk Member
Seaford Soft Chalk Member
Lewes Nodular Chalk Member
Ranscombe Griotte Chalk Member

This lithostratigraphical scheme, which is used here (following Mortimore, 1986a), comprises a succession of distinctive chalk lithologies, each reflected by its own geophysical signature. The Ranscombe Member, formerly known as the Caburn Member (Mortimore, 1983), is characterised by both types of griotte chalk, with sparse nodular flints and horizons of true nodular chalk in the highest beds; the upper part of the sequence contains numerous discrete marl seams. The post-Ranscombe succession is flinty throughout, in contrast to the essentially flintless Ranscombe Member. The Lewes Member comprises nodular incipient hardground chalks. The Seaford Member is characterised by soft white chalk with many courses of nodular flints (some very large), and numerous iron-stained spongiferous horizons representing minor sedimentary breaks; it thus represents a marked

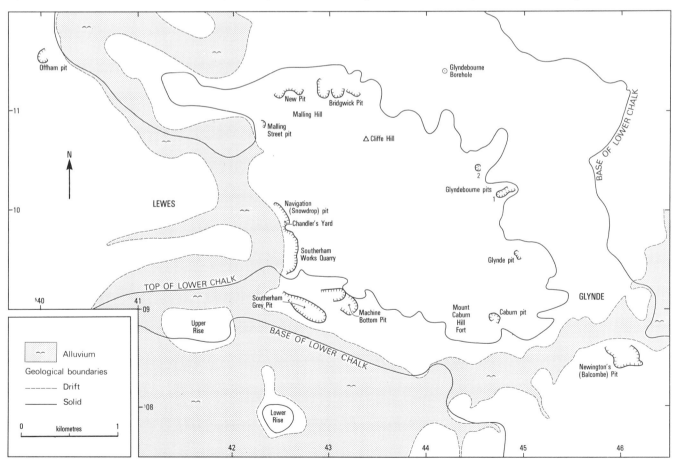

Figure 13 Sketch-map showing localities in the Mount Caburn area

change in the depositional environment. Marly griotte chalk with discrete marl seams returns in the Newhaven Member, and the overlying essentially marl-free Culver Member is typified by soft chalks with nodular flints. Marl seams are found again in the Portsdown Member. In this sequence the correlative of the Basal Complex discussed above falls within the Lewes Member.

Using geophysical logs, particularly 16-inch resistivity and natural gamma, it is possible to use the scheme to classify the strata in uncored wells and boreholes, and to study lateral variation in the subcrop succession.

The Lewes district is of particular importance in Chalk lithostratigraphy in that it has provided many of the strato-types of members, beds and marker-horizons in the new scheme. The choice of Lewes was not fortuitous, but was determined by the fact that the thickest and most complete successions in southern England are found within a clearly-defined Sussex depositional trough with its axis passing through the present district (see Mortimore, 1983, figs. 3a–d). The relationship between the members, main litho-stratigraphical marker-horizons, conventional biozones and international stages is shown in Figure 12. The relative stratigraphical positions of the most important sections in the district are also shown here. The locations of exposures in the Lewes area are shown in Figure 13. The distribution of the members of the Upper and Middle Chalk is shown in Figure 14.   CJW,RNM

BIOSTRATIGRAPHICAL SUBDIVISION

The Chalk is currently zoned on the basis of macrofaunal assemblages using brachiopods, bivalves, ammonites, belemnites and crinoids as zonal indices, although in recent years complementary zonal schemes have been erected using dominantly benthonic forms of foraminifera for the Cenomanian (zones 7 to 14 of Carter and Hart, 1977) and for the Coniacian to Maastrichtian stages (Bailey and others, 1983). Modern research has also demonstrated the utility for correlation purposes of faunal marker bands which reflect biostratigraphical 'events'.

Lower Chalk

GLAUCONITIC MARL

The Glauconitic Marl, the basal bed of the Lower Chalk, is the 'Chloritic Marl' of early authors. It consists of sandy, locally slightly micaceous, bioturbated marl with abundant glauconite grains and scattered dark grey phosphatic nodules. It is up to 1.8 m thick. The bed is seldom exposed but has been penetrated by a number of cored boreholes in which it rests on the Gault with a sharp, commonly bur-rowed contact. At Rodmell [441 071] a slight unconformity has been noted beneath this bed (p. 66). The Glauconitic Marl grades upwards into the Chalk Marl, and hence its top can only be defined arbitrarily.

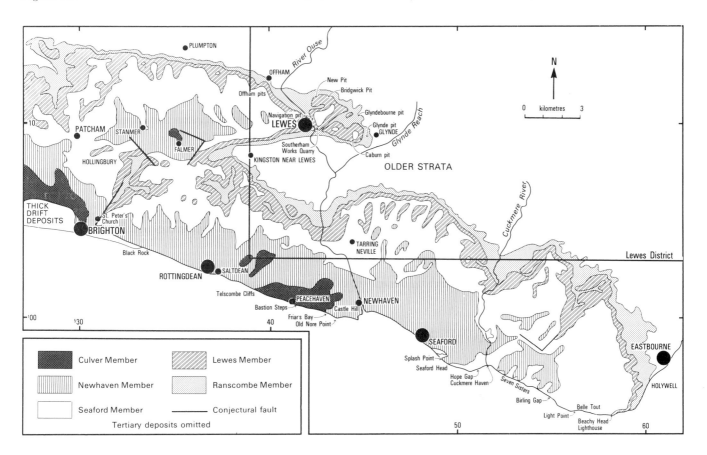

Figure 14 Sketch-map showing the distribution of the members of the Upper and Middle Chalk; derived from the zonal maps of Gaster (1939; 1951)

Fossils are typically few and poorly preserved; those recovered are characterised by a bivalve assemblage of small *Aucellina*, including *A. gryphaeoides* and *A. uerpmanni* (see Morter and Wood, 1983). This bed corresponds to the benthonic foraminiferal Zone 7. The above assemblage also occurs in the succeeding (about 7 m) beds (of Zone 8) at Folkestone and the combined succession probably equates with the *Neostlingoceras* [*Hypoturrilites*] *carcitanense* Subzone of the *Mantelliceras mantelli* Zone. It should be noted that at Eastbourne the Glauconitic Marl was assigned to Zone 9: this anomaly cannot be resolved at present (Figure 15).

CHALK MARL

The Chalk Marl comprises a rhythmically interbedded sequence of grey marls and hard sponge-bearing marly limestones from 22 to over 40 m thick. In fresh borehole cores the lithologies are very dark grey, but become paler grey and greyish white on drying; nodules of radiating crystalline pyrite are locally common. The rhythms are best studied in borehole cores, and details of the BGS boreholes are retained on file.

In its simplest form, each rhythm resembles those described by Destombes and Shephard-Thorn (1971, fig. 2b) and consists of very marly dark grey chalk with abundant foraminifera passing upwards into less marly paler grey chalk, commonly terminated by a very hard sponge-bearing calcisphere limestone with an eroded upper surface. Individual rhythms are usually more complex and many are incomplete. In places harder beds appear to have formed by diagenetic segregation of calcium carbonate. Some of these hard beds are very fossiliferous, yielding uncrushed ammonites, bivalves, brachiopods, and sponges. Fossils are also abundant in the more marly beds but are generally crushed and poorly preserved. Large exposures in the Chalk Marl can be seen at Southerham Grey Pit and at Glynde.

The Chalk Marl comprises the upper part of the *Mantelliceras mantelli* Zone and the *M. dixoni* Zone of the Lower Cenomanian and the lower part of the *Acanthoceras rhotomagense* Zone of the Middle Cenomanian (Figure 12). Within this sequence there are several key macrofossil bands which are as follows in ascending order (see also p. 68):

1 The total range of *Monticlarella? rectifrons*. This species is tentatively assigned to *Monticlarella* on its external characters, but lacks the distinctive laterally 'crinkled' anterior commissure of that genus. *M. rectifrons* is relatively common in the Lower Cenomanian Chalk Marl of southern England, but has been commonly misidentified as the superficially similar Middle Cenomanian species *Grasirhynchia martini*. It was this confusion that led to the establishment of a '*Rhynchonella martini* Zone' (subsequently replaced by the *Schloenbachia varians* Zone) for the Chalk Marl (e.g. Penning and Jukes-Browne, 1881). The upper part of the range of this critical species overlaps with the band of *Inoceramus* ex gr. *virgatus* (see below). The total range appears to include the *Mantelliceras saxbii* Subzone with an acme at the top of the subzone, and extends up into the lower part of the *M. dixoni* Zone.

2 A band characterised by an assemblage of thick-shelled inflated relatively weakly sculptured inoceramids belonging to the group of *Inoceramus virgatus*. All the inoceramids of this assemblage would previously have been referred to the nomenclaturally invalid taxon *I. etheridgei*, for which Böhm (1915) proposed the replacement name

I. scalprum. Sornay (1978) has shown that the inoceramids of the '*etheridgei*' assemblage should be referred to the two subspecies *I. virgatus virgatus* and *I. virgatus scalprum*, corresponding to the more quadrangular elongate and the more rounded forms respectively. This band is a very important stratigraphical marker, normally occurring at the top of a succession characterised by spongiferous limestones with *Exanthesis*.

3 The lower *Orbirhynchia mantelliana* band, characterised by common *O. mantelliana*, particularly at the base. This band coincides with the first appearance of the Middle Cenomanian ammonite genus *Acanthoceras* at Southerham (Figure 16) and elsewhere, and its top is approximately indicated by the lower two of three particularly prominent limestones.

4 A band of usually dark-coloured, silty, highly fossiliferous chalk with a diverse fauna of small brachiopods comprising *Grasirhynchia martini*, *Kingena concinna*, *Modestella geinitzi* and *Platythyris squamosa*. This band occupies a position between the two *Orbirhynchia mantelliana* bands and can be traced not only throughout southern England but also to the coastal sections of the Boulonnais in northern France. Its base occurs less than 1 m above the uppermost of the three prominent limestones at Southerham (Figure 16). The brachiopod assemblage is characterised by *Modestella geinitzi*, which appears to be restricted to this level, and by *Grasirhychia martini*, which ranges up to the upper *Orbirhynchia mantelliana* band. This brachiopod band overlies a bed rich in the bivalve *Entolium sp.* (*E. laminosum?*) with *Oxytoma seminudum*, the base of which approximates to the boundary between benthonic foraminiferal zones 10 and 11i, taken by Carter and Hart (1977) to mark the base of the Middle Cenomanian.

5 The upper *Orbirhynchia mantelliana* band (equivalent to the *O. mantelliana* Band of Kennedy, 1969). This band is characterised by common *O. mantelliana*, with abundant *Sciponoceras* and other heteromorph ammonites at the top. In Sussex the upper limit of the band occurs only a few metres beneath the lithological change from rhythmic Chalk Marl to massive Grey Chalk and coincides with the so-called 'mid-Cenomanian non-sequence' (Carter and Hart, 1977). The latter is marked by a dramatic change in the foraminiferal assemblages from predominantly benthonic below to predominantly planktonic above, a change interpreted by Carter and Hart as indicative of a deepening of the sea.

Using a combination of these macrofossil markers and published and unpublished data on the limits of the benthonic foraminiferal zones in the Aycliff (Dover) and Glyndebourne boreholes, it has been possible to establish a satisfactory outline correlation between these boreholes and the thinner succession in the Rodmill Borehole [6008 0070] (Figure 15). Correlation between the Rodmill succession and the even further reduced succession at Beachy Head presents difficulties, however, and only tentative tie-lines are shown in the figure for this reason. The succession below the *Inoceramus* ex gr. *virgatus* band in the Glyndebourne Borehole is only some 3 m thicker than that in the Aycliff Borehole, but the stratigraphy interpreted from the ranges and/or positions of the key biostratigraphical marker bands (in common with successions elsewhere in Sussex) is significantly different. The known vertical range of *Aucellina* in the Glyndebourne Borehole is restricted to a single occurrence of finely striate *A. gryphaeoides* within the Chalk Marl, and there are no records from the basal 3 m of Chalk Marl below the lowest occurrence of *Monticlarella? rectifrons*. There is a concomitant reduction in thickness of benthonic foraminiferal Zone 8 (Zone 7 not being recognised in this borehole) and a doubling in thickness of Zone 9, reflected in the greatly increased range of *M? rectifrons*. The succession above the *I.* ex gr.

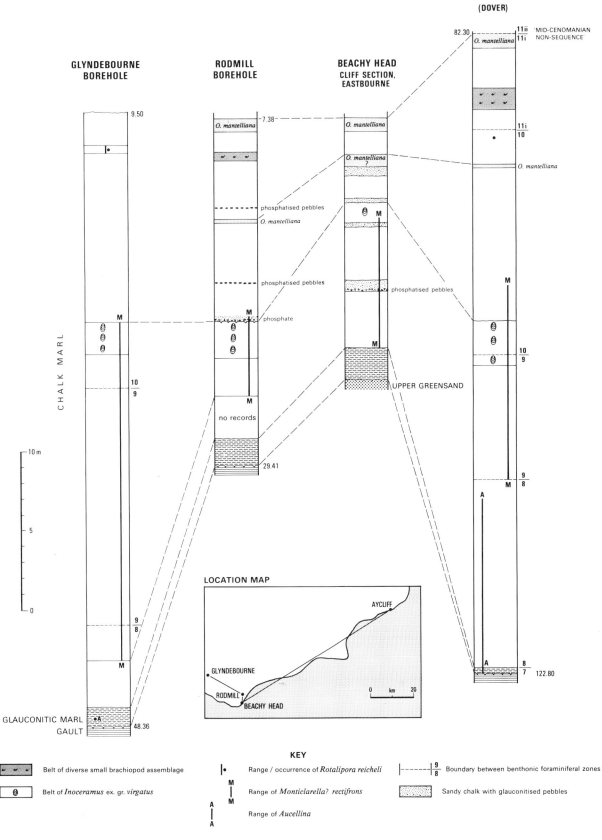

Figure 15   Outline correlation of the lower part of the Lower Chalk of the Lewes–Eastbourne area with that of Dover

virgatus band probably reaches virtually to the base of the Middle Cenomanian (base of Zone 11i) as defined by benthonic foraminifera. The evidence for this is provided by a flood of the planktonic foraminifer *Rotalipora reicheli* between 11.5 and 12 m below the top of the borehole, comparable with the flood that typically occurs just beneath the base of Zone 11i (Professor M. B. Hart, personal communication) and approximates to the entry of *Acanthoceras* (see Amedro and others, 1978).

The Rodmill Borehole shows a much thinner succession than that of Glyndebourne (Figure 15). In Rodmill, a bed of sparsely glauconitic chalk with glauconitised clasts and phosphatised pebbles overlies the *I.* ex gr. *virgatus* band, and an horizon of phosphatised pebbles is found in the interval between the latter and the lower *Orbirhynchia mantelliana* band. Another horizon with phosphatised pebbles is present some 0.7 m above the lower *O. mantelliana* band, and *Oxytoma seminudum* was collected from 'gritty' chalk immediately below. The three Middle Cenomanian brachiopod bands are readily identifiable, and exhibit only a small degree of condensation relative to Southerham Grey Pit and Aycliff (6.5 m as against 8.5 m). The main thickness reduction occurs in the succession below the *I.* ex gr. *virgatus* band, and this is reflected in the reduced proven vertical range of *M.? rectifrons*.

Further condensation is seen in the Eastbourne section at Falling Sands (Figure 15), with the presence of several beds of gritty chalk with glauconitised pebbles. This section is extensively faulted and is usually indifferently exposed. Measurements by Miss H. Anderson (formerly of Brighton Polytechnic) agree well with those of Kennedy (1969, p.507), apart from a thickness of 1.7 m not present in his section and presumably not then exposed. This missing part of the succession (included in Figure 15), comprises a further pebble bed intercalated between two conspicuous marls and overlain by a limestone. In the lower part of the Chalk Marl, a 0.7-m bed of chalk rich in granular glauconite and pelletal phosphate yields a rich *Mantelliceras saxbii* Subzone ammonite fauna preserved both as (derived) phosphatised and glauconitised steinkerns, and as weakly glauconitised chalk moulds (see Kennedy, 1969, p.507 and table 8). The biostratigraphy of the succession above the *saxbii* pebble bed is imperfectly known, and consequently only tentative correlations are indicated in Figure 15.

The Southerham Grey Pit succession (Figure 16) is comparable in thickness to that of Aycliff, and the combined thicknesses of the various Middle Cenomanian brachiopod bands are almost identical. Reduction has occurred at Southerham relative to Aycliff in the beds below the *Inoceramus* ex gr. *virgatus* band. On the other hand, relative expansion in the Southerham succession is seen in the interval between the *I.* ex gr. *virgatus* band and the lower *Orbirhynchia mantelliana* band. Compared with the corresponding part of the Southerham succession, the beds below the *I.* ex gr. *virgatus* band in Glyndebourne appear to be somewhat expanded (20 m; 26 m respectively).

GREY CHALK

No formal definition has ever been given for the Grey Chalk; the term is used here to include all the Chalk between the top of the Chalk Marl and the base of the Plenus Marls. This subdivision displays much less variation in thickness in this district than does the Chalk Marl and is typically about 35 m thick.

The Chalk Marl passes gradationally upwards into the Grey Chalk, although locally, as for example at Southerham Grey Pit (Plate 6), the transition is relatively abrupt and occurs about 1 m above the upper *O. mantelliana* band (p. 50). This chalk is typically less argillaceous and paler grey than the Chalk Marl, with less definite rhythms, and with thicker units of hard, pale, and less marly chalk (see also Destombes and Shephard-Thorn, 1971). In the middle of this sequence, a distinctive massive bed with 'laminated structures' (scours) occurs at Southerham (Grey Pit and Machine Bottom Pit) and Beachy Head (Kennedy, 1969; see also Figure 16). This is the equivalent of Bed 7 of the Folkestone section (Jukes-Browne and Hill, 1903), but correlation at this level locally presents problems.

The Grey Chalk comprises the upper part of the *rhotomagense* Zone and the whole of the *Acanthoceras jukesbrownei* Zone of the Middle Cenomanian, together with the Upper Cenomanian *Calycoceras guerangeri* Zone. The biostratigraphy is imperfectly known in this district (but see p. 68).

PLENUS MARLS

The Plenus Marls consist largely of soft marly chalks with a characteristic very slight greenish grey tinge; they are up to 11 m thick (Figure 17). Jefferies (1961; 1963), in a detailed study of the microfauna and macrofauna of the Plenus Marls over a wide area of the Anglo-Paris Basin, has demonstrated the lateral persistence of individual beds. He recognised a succession of eight beds with major erosion surfaces both within the sequence and at its base. These beds are coextensive with the zone of *Metoicoceras geslinianum*. The characteristic belemnite *Actinocamax plenus* is rare in this district and the marls are generally poorly fossiliferous.

Within the Plenus Marls there are three more or less distinct erosion surfaces in addition to the basal burrowed surface, at the tops of Jefferies's beds 1, 3 and 5 respectively. Faunal changes occur at these levels. From regional studies it is apparent that Bed 1 is characterised by abundant gryphaeate oysters (*Pycnodonte*) and the rhynchonellid brachiopod *Orbirhynchia multicostata*. The only indication of originally aragonite-shelled macrofossils are sparse oyster casts. The terminal erosion surface (of Bed 1) coincides with the upper limit of the planktonic foraminifer *Rotalipora greenhornensis*. Above this erosion surface, aragonitic molluscs such as ammonites, gastropods and bivalves are relatively common, particularly in the thicker successions. These fossils are preserved either as limonite- or green-coated moulds, according to whether the original shell was converted to pyrite or to calcite.

The erosion surface at the top of Bed 3 marks a major change in both the macrofossil and planktonic foraminiferal assemblages. The surface coincides with the extinction-datum of *Rotalipora cushmani*, a datum regarded by micropalaeontologists as globally isochronous, and formerly taken to mark the top of the Cenomanian Stage. The overlying beds 4 to 6 are characterised by a macrofossil assemblage comprising *Actinocamax plenus*, '*Aequipecten*' *arlesiensis* and

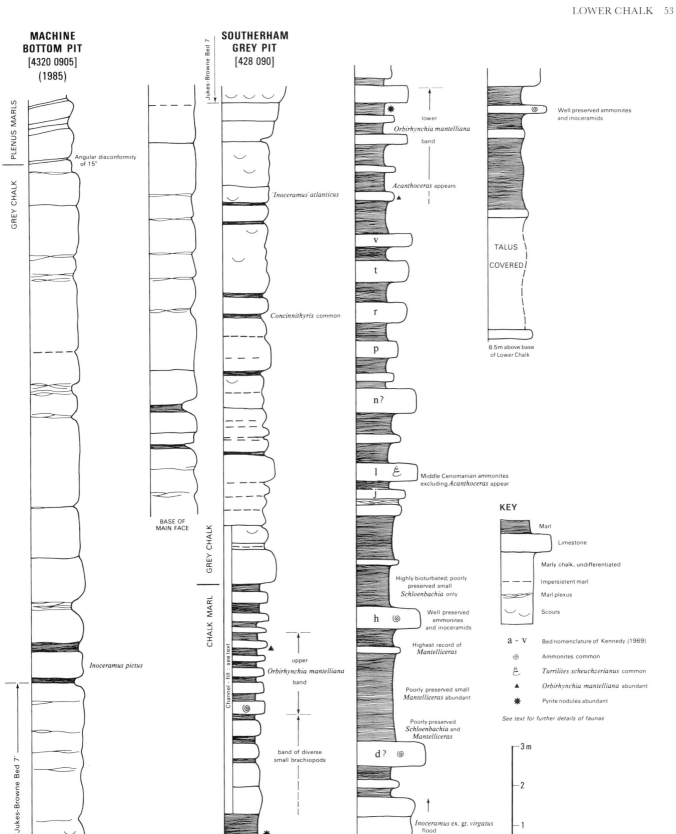

Figure 16   The Lower Chalk succession at Southerham

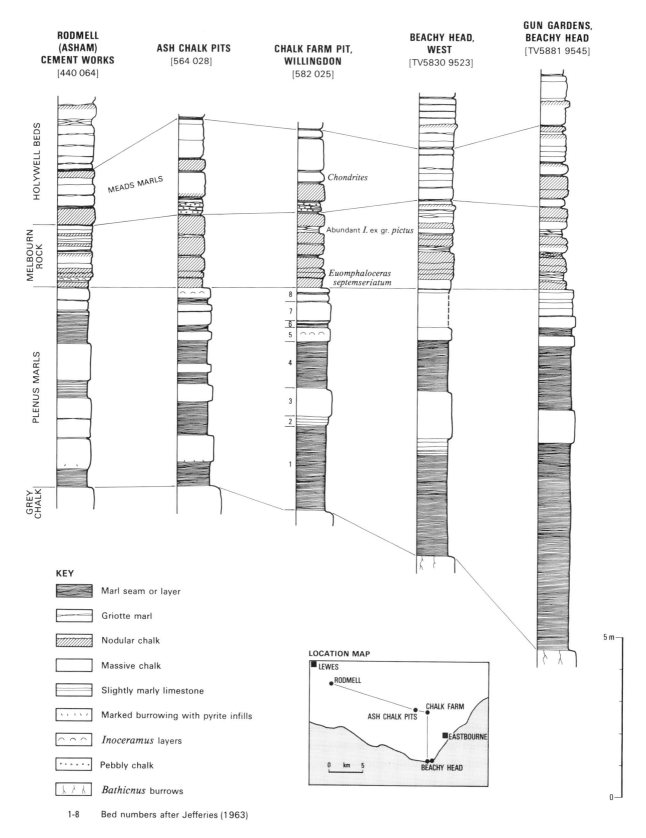

KEY

Marl seam or layer

Griotte marl

Nodular chalk

Massive chalk

Slightly marly limestone

Marked burrowing with pyrite infills

Inoceramus layers

Pebbly chalk

Bathicnus burrows

1-8   Bed numbers after Jefferies (1963)

Figure 17   Sections in the Plenus Marls and Melbourn Rock of the Lewes and Eastbourne districts

Oxytoma seminudum, and were believed by Jefferies (1961) to represent an incursion from an area of cooler water to the north during a period of lowered sea temperatures. *Actinocamax plenus* itself thus does not occur throughout the lithostratigraphical unit to which it gives its name, but is restricted to beds 4 to 6 (in which it is locally common), apart from extremely rare occurrences at the top of Bed 3, and some questionable records in beds 7 and 8.

There are also significant changes in the ammonite assemblages within the Plenus Marls, particularly at the terminal erosion surface of Bed 3. Beds 1 to 3 yield *Calycoceras naviculare*, *Metoicoceras geslinianum*, *Puzosia (Anapuzosia) dibleyi*, and a number of rare species cited by Wright and Kennedy (1981). *C. naviculare* is more common in the lower part of the Plenus Marls than it is in the underlying *guerangeri* (formerly *naviculare*) Zone, and the holotype, stated to be from the 'Upper Chalk of Offham', almost certainly came from this part of the succession. The ammonites of beds 4 to 8 comprise, in addition to *M. geslinianum*, *Euomphaloceras septemseriatum*, *Pseudocalycoceras dentonense* and *Sciponoceras gracile*, an assemblage which correlates with the late Cenomanian *Sciponoceras gracile* Zone of North America. *M. geslinianum* ranges up to Bed 7 although, in common with other ammonites, it has not been recorded from Bed 6. Of the remaining species, *E. septemseriatum* is first found in Bed 7, and ranges upwards, whereas *P. dentonense* is apparently restricted to Bed 5. The Plenus Marls are particularly characterised by the small brachiopod *Monticlarella jefferiesi* (cited by Jefferies (1961) as '*Rhynchonella*' *lineolata* Philips *carteri* Davidson), and *Orbirhynchia wiesti* is restricted to Bed 7, in which it is common. Full details of the remaining macrofossils and microfossils of the Plenus Marls are to be found in Jefferies (1961).

In the Lewes district, there are only limited faunal records from the Plenus Marls. The best section is at Chalk Farm [582 025] (Figure 17), outside the Lewes district, and there the lower beds are not particularly accessible. *Actinocamax plenus* has been collected both from Chalk Farm and Rodmell in Bed 4, together with fragments of *Metoicoceras*. Bed 6 at Chalk Farm is rich in inoceramids including *Inoceramus pictus* and an irregularly rugate form possibly referable to *I. ginterensis*. The most fossiliferous section, and the one from which ammonites can most easily be collected, is Beachy Head in the Eastbourne district. Of particular interest here is the presence of the elongate burrow tracefossil *Bathichnus paramoudrae* (Bromley and others, 1975), both in the basal part of the Plenus Marls, and extending down into the *guerangeri* Zone chalk from beneath the basal burrowed surface. Higher in the Chalk, *B. paramoudrae* is associated with the giant vertical cylindrical flints known as paramoudras, but at this stratigraphical level flints are absent, and all that can be seen is an annular zone of discolouration several centimetres across, around the central burrow, which is itself some 2 to 3 mm in diameter.

Upper and Middle Chalk

RANSCOMBE GRIOTTE CHALK MEMBER

The Ranscombe Member comprises essentially flintless griotte chalk, with discrete marl seams and (locally) mature mineralised hardgrounds. A composite total thickness of 80 m has been compiled from the Mount Caburn sections. The lower limit of the member is drawn at the base of the Melbourn Rock (Figure 12) and the upper limit is taken at the erosion surface beneath the lower Glynde Marl (p. 59). The name is taken from Ranscombe Lane, Mount Caburn, and the stratotype is the Caburn pit [4473 0891]; the basal boundary stratotype is, however, taken at Holywell near Eastbourne, as the basal beds of the member are no longer exposed at Mount Caburn.

The Ranscombe Member of Mortimore (1983) does not equate with the 'Marnes de Ranscombe' used by Barrois (1876) from a now-vanished pit farther along Ranscombe Lane; Barrois's unit approximately corresponds to the upper part of the Ranscombe Member (the New Pit Beds), together with the lower part of the overlying Lewes Member up to the traditional base of the Upper Chalk.

Three lithological units can be recognised within the Ranscombe Member; they are, in ascending order, the Melbourn Rock, the Holywell Beds and the New Pit Beds (see Figure 12).

The Melbourn Rock comprises 2 to 3 m of closely-spaced beds of typically pale reddish grey (iron-stained) nodular griotte chalk alternating with smooth griotte chalk with intercalated marl seams. Within the district it does not form such a well-marked rock as it does at the type locality in southern Cambridgeshire, but the relatively indurated nodular beds contrast strongly with the underlying Plenus Marls, and with the succeeding smooth griotte chalks. The relationship between the Melbourn Rock of Sussex and that of the type area is unclear, but the base of the unit is believed to be the same in both areas. The upper limit of the nodularity is variable, but the top of the Melbourn Rock, as used here, can be defined conveniently by the base of the lowest marl of three conspicuous closely-spaced pairs of marls (the Meads Marls) in about the lowest 2 m of the overlying Holywell Beds.

The Holywell Beds, named after Holywell near Eastbourne, are about 40 m thick and may be conveniently described in three parts of comparable thickness (Figure 18). The lower Holywell Beds mainly comprise smooth griotte chalks. Apart from the three marl-pairs in the basal part (Meads Marls), there are an additional four more or less well developed griotte marls (the Holywell Marls) higher in the sequence, of which the lowest is particularly conspicuous. Small flints were reported from the lower Holywell Beds just above the Melbourn Rock near Milton Street [5358 0360] (White, 1926, p.52; Gaster, 1939). The middle Holywell Beds are characterised by chalks gritty with inoceramid debris (*Mytiloides sp.*), including both fragmented and complete shells. There are four griotte marls (the Gun Garden Marls), of which the uppermost is best developed, and several red iron-stained nodular beds. In the upper Holywell Beds there is a return to more massive chalks without nodular beds and with more widely spaced and weakly developed griotte marls.

The Holywell Beds are succeeded by a succession of smooth griotte chalks, 40 to 45 m thick, with discrete griotte marls at irregular intervals: these are the New Pit Beds, named after the New Pit, Malling Hill [4249 1120], now occupied by a council depot. The marl seams within the New Pit Beds can be divided into three groups of two, each pair

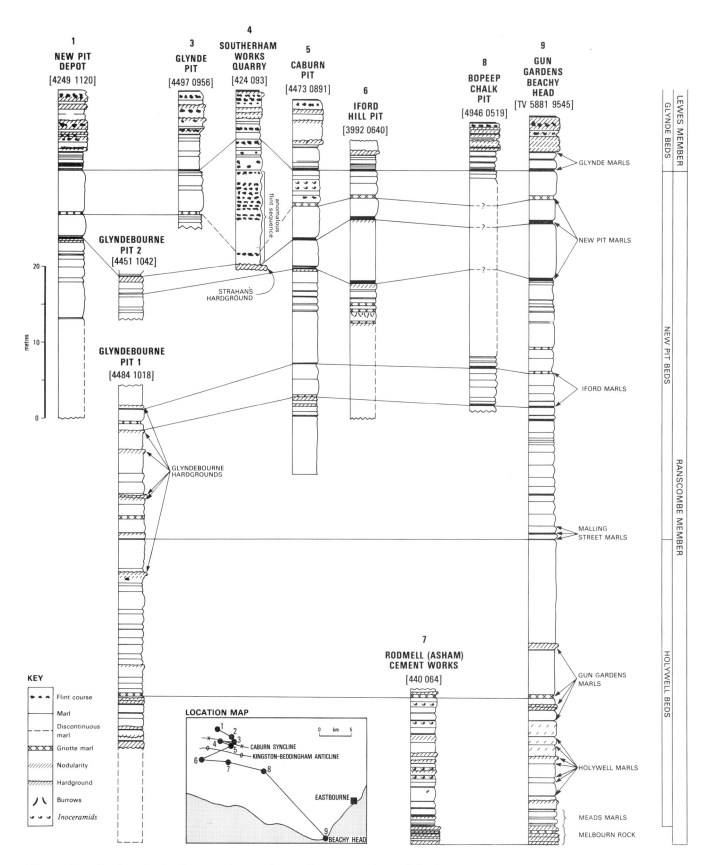

Figure 18   Sections in the Ranscombe and lowest Lewes members of the Lewes and Eastbourne districts

providing a distinctive signature in geophysical logs (see p. 49 and Figure 18). At the base are found the Malling Street Marls, two discrete 3 to 5-cm plastic marl seams named after the pits beside Malling Street [423 109], which were figured by Mantell (1822, fig. 3). The Iford Marls, named after the disused pit east of Iford Hill [3992 0640], are well developed in places, for example at the type locality, but elsewhere are not apparent at outcrop, despite the fact that they can be recognised in geophysical logs. The New Pit Marls 1 and 2, towards the top of the succession, are much thicker and more conspicuous than the other marl seams of the New Pit Beds, and they produce a very distinctive pair of resistivity minimum 'spikes' that can be traced throughout southern England in geophysical logs.

In the area of Mount Caburn, there are two anomalous successions within the higher part of the Ranscombe Member. The Glyndebourne pit [4484 1018] exposes a sequence of six mineralised hardgrounds, the Glyndebourne Hardgrounds, first described by White (1926, p.52 and fig. 7) (see p. 72 and Figure 18). Of these hardgrounds, the lowest (Glyndebourne 1), is situated below the paired Malling Street Marls, and thus falls within the upper part of the Holywell Beds; the remaining five hardgrounds represent a condensation of part of the New Pit Beds. The tentative correlation between this and nearby sequences is shown in Figure 18. In the Southerham Works Quarry [4265 0965] there is a thin and very localised phosphatic chalk overlying a massive 70-cm-thick mineralised hardground, named Strahan's Hardground after its discoverer (Strahan, 1896). Strahan considered that the phosphatic chalk correlated with the Chalk Rock, but geophysical evidence indicates that Strahan's Hardground and associated phosphatic chalk equate with the higher of the New Pit Marls (New Pit Marl 2) near the top of the New Pit Beds (Figure 18). Above this hardground there occurs an anomalous sequence of small paramoudra-like flints discussed by Mortimore (1986b).

The lowest recorded fauna from the Ranscombe Member is that from a thin bed about 30 cm above the base of the Melbourn Rock, which includes abundant *Sciponoceras bohemicum anterius* and subordinate *S. gracile*, and thus provides evidence for the *Neocardioceras juddii* Zone at the top of the Cenomanian. At Chalk Farm (Figure 17) small extremely poorly preserved pebble-fossils of *Euomphaloceras septemseriatum* have been collected from this horizon, and the specimen of *Allocrioceras annulatum* from Eastbourne (BGS GSM 87749: cited by Kennedy (1967) as *Stomohamites sp.*) is also believed to come from this level. Immediately beneath the top of the Melbourn Rock, a shell-bed has yielded inoceramids of Cenomanian aspect at Chalk Farm, including *Inoceramus* ex gr. *pictus* and another, possibly unrelated form. There is so far no evidence of the inoceramid genus *Mytiloides* from the Melbourn Rock as defined here.

Evidence of the basal Turonian *Watinoceras coloradoense* Zone of the international ammonite zonation is provided by a poorly preserved *Watinoceras* cf. *amudariense* from 3.4 m above the base of the Melbourn Rock at Holywell, Eastbourne (Wright and Kennedy, 1981, pl.5, fig. 5). According to the present definition, this horizon lies above the Melbourn Rock, within the unit with three closely-spaced marl-pairs (Meads Marls) at the base of the Holywell Beds. On the limited evidence from these beds and from the

Melbourn Rock, the base of the Turonian Stage should be drawn at or near the base of the Holywell Beds. These results are, however, somewhat at variance with collections made by Dr P. Woodroof (preserved in the Oxford University Museum). Woodroof's material from a presumed equivalent horizon to that which yielded the *Watinoceras* (just above the middle of the three marl-pairs) comprises an assemblage of late Cenomanian aspect including *I.* ex gr. *pictus* and *I. pictus bohemicus*, together with four specimens of *Sciponoceras sp.* and an ammonite identified by Dr W. J. Kennedy as *Thomelites* cf. *serotinus*.

Comparatively little biostratigraphical information is available at present for the higher part of the Holywell Beds. *Mammites* occurs between Holywell Marls 2 and 3 and *Lewesiceras* from above Holywell Marl 3. The Woodroof collection includes *Mytiloides sp.* and *Inoceramus apicalis* from nodular chalks about 2 m beneath the lowest of the Gun Garden Marls. *Lewesiceras* occurs beneath Gun Gardens Marl 2. In the middle part of the Holywell Beds, between the third and main (fourth) Gun Garden Marl, *Mytiloides mytiloides* occurs in flood abundance as whole and fragmented shells. These inoceramids are associated with *Orbirhynchia cuvieri* and *Conulus subrotundus*. *Mammites nodosoides* has been collected from this part of the succession near the base of the main Glyndebourne pit. The uppermost approximately 3 m of chalk beneath the Gun Gardens Main Marl yields *Mytiloides* with an inflated umbo, which are possibly transitional between the flat *M. mytiloides* and the highly inflated *M. labiatus*. *Orbirhynchia sp.*, *Concinnithyris sp.* and *Conulus subrotundus* are relatively common 1.5 m beneath the Gun Gardens Main Marl at Eastbourne. The highest part of the Holywell Beds is relatively poorly fossiliferous but *Orbirhynchia sp.* and *Concinnithyris sp.* occur at Eastbourne, and there are records of *Mytiloides sp.* from Glyndebourne between the Glyndebourne Hardground 1 and the Malling Street Marls.

The Malling Street Marls mark the entry point of the small brachiopod *Terebratulina lata*, and can thus be taken as the base of the traditional *T. lata* Zone. There are several ammonite records of great importance for long-range correlation. *Collignoceras woolgari*, the zonal index for the Middle Turonian in the current tripartite ammonite zonal scheme, was originally described from the Lewes area. This species is not uncommon just above the Malling Street Marls (e.g. in the Glyndebourne pit), and it has also been collected from the nodular chalk underlying the New Pit Marls in the Glyndebourne small pit [4451 1042] (Figure 19). The Glyndebourne Hardgrounds 2 and 3 in the anomalous succession in the Glyndebourne pit have yielded a small *Lewesiceras peramplum* and a probably undescribed *Scaphites*. *Romaniceras* (*Yubariceras*) *ornatissimum*, another long-ranging ammonite in the Middle Turonian, is known from a single specimen (BGS GSM117334) in the Malling Pit from the New Pit Marl 2 – Glynde Marls interval. The New Pit Beds mark a major change in the inoceramids, with the replacement of the *Mytiloides*-dominated assemblages of the underlying Holywell Beds by assemblages dominated by *Inoceramus spp.*, including *I. apicalis*, *I. cuvierii* s.s. and rugate morphotypes of *I. cuvierii*. These inoceramids range throughout the New Pit Beds, but do not become common until above the New Pit Marls. White (1926) recorded *Inoceramus labiatus*

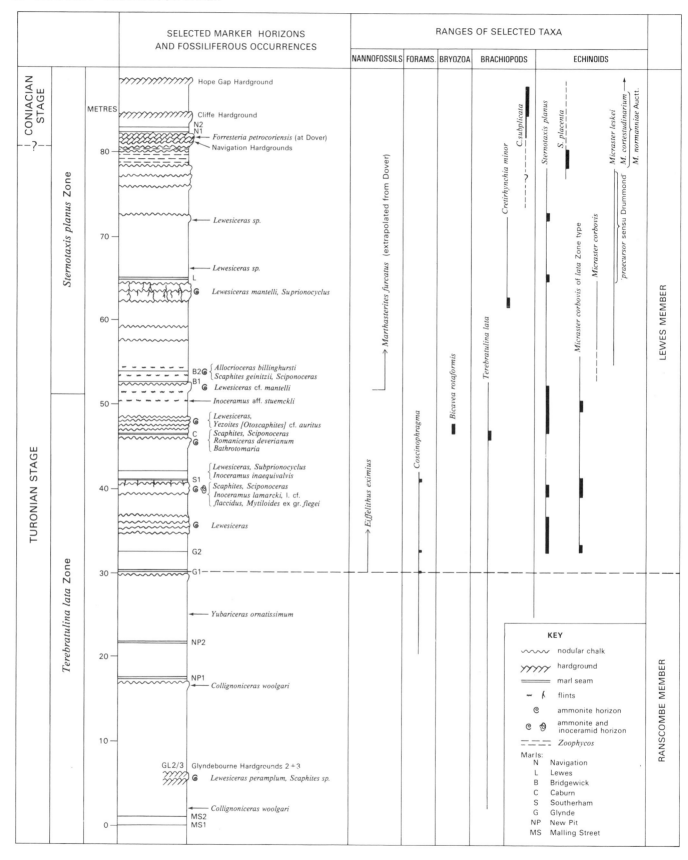

Figure 19   The biostratigraphy of the *lata* and *planus* zones

(i.e. *Mytiloides sp.*) as ranging up to the Glyndebourne Hardgrounds 2 and 3, but this observation has not been confirmed.

Apart from inoceramids and sparse ammonites, the fauna of the New Pit Beds is of low diversity. Rhynchonellid (*Orbirhynchia*) and terebratulid (*Concinnithyris*) brachiopods occur, but there is little detailed information on their distribution, or on which species are represented. There is a well-developed *Conulus* band in the New Pit Marls–Glynde Marls interval, with *Conulus subrotundus*. This constitutes the highest record of the genus until its reappearance in the (Coniacian) Seaford Member. The interval between the paired New Pit Marls has yielded the lowest occurrence of the large complex chambered lituolid foraminifer *Coscinophragma*. This distinctive microfossil is taken by some workers to indicate a relatively deep-water environment.

Lewes Nodular Chalk Member

The Lewes Member, 58 to 89 m thick, comprises a complex succession of nodular and hardground chalks with discrete marl seams particularly in the lower part, and closely-spaced courses of nodular thalassinoidean burrow-form flints. The lower limit is taken at the base of the lower Glynde Marl; in general this dark, plastic marl seam closely marks the base of flinty chalk within this district. The upper limit is taken at an erosion-surface immediately above the upper of the paired Shoreham Marls (Figure 12). The stratotype is composite, taken from the exposures around the town of Lewes, particularly those around the perimeter of Mount Caburn; the basal boundary stratotype is the Caburn pit [4473 0891], which provides a magnificent air-weathered section through the junction between the Ranscombe Member and the Lewes Member (Figures 18 and 20).

Griotte texture like that of the Ranscombe Member is present in the softer chalks between the hardgrounds and the nodular chalks in the lower part of the succession, but above the Lewes Nodular Chalks (Figure 12) this texture is replaced by Bänderkreide or *Zoophycos* chalk. This term describes chalk characterised by the complex helical trace-fossil *Zoophycos* (Voigt and Häntzschel, 1956; Ekdale and Bromley, 1984; Bromley and Ekdale, 1984) which appears in section as discrete slivers of darker chalk within a paler matrix. Bänkerkreide persists in the softer chalk right up to the top of the member but is particularly well developed above the Lewes Nodular Chalks and near the top of the member in the Beachy Head Sponge Beds.

The lower part of the Lewes Member, up to the Navigation Hardgrounds (see Figure 12), is divided into six units (beds) each of which, except for the uppermost (Navigation Beds), has a major marl seam at its base. The Glynde Marl 1—at the base of the Glynde Beds, 10.7 m thick at the now largely obscured pit [4497 0956] 750 m WNW of Glynde church—is dark-coloured with a plastic 'buttery' texture. It is overlain by a sequence of marly chalks, above which the first regular courses of nodular flints occur within the nodular chalks (Figures 18 and 20). There are two bands of nodular chalks separated by griotte marls of which the upper yields moulds of ammonites and is separated from the Southerham Marl 1 (at the top of the Glynde Beds) by a band of small digitate thalassinoidean burrow-form flints im-

mediately beneath the marl and which is underlain by larger nodular flints.

The paired Southerham Marls are the equivalent of the paired marls delimiting the so-called '4-ft-band' of the Dover succession (Figure 20; Rowe, 1900). The lower marl is dark and plastic, whereas the marls higher in the Lewes Member are of a brittle type. The lower marl is also characterised by an acme occurrence of the foraminifera *Coscinophragma sp.*, a fossil which is clearly visible to the naked eye in a hand-specimen. The associated occurrence of a marl with *Coscinophragma* and underlying digitate burrow-form flints can be traced throughout southern England and into East Anglia (Mortimore and Wood, 1986), thus providing a lithological marker for long-range correlation. The Caburn and Ringmer beds (5.2 m and 6 to 8 m thick respectively) each comprise a lower and upper sequence of nodular chalks separated by a griotte unit. The upper nodular chalk unit in the Caburn Beds is strongly indurated, and yields sponges together with sparse ammonites; small nodular flints are present above each sponge bed. At a corresponding level in the Ringmer Beds two courses of large nodular flints, the Bridgewick Flints, are found within indurated nodular chalks.

The Bridgewick Flints are overlain by two (locally three) well-developed discrete marl seams, the Bridgewick Marls, which are named, like the Bridgewick Flints, after the earlier spelling of the Bridgwick pit [4310 1117] on the north side of Mount Caburn. Between and above the paired Bridgewick Marls are found courses of large nodular flints, the Bopeep Flints, named after the Bopeep pit [4946 0519]. The combination of Bridgewick Flints, Bopeep Flints and intercalated Bridgewick Marls equates with the 'Basal Complex' at Dover (see p. 47 and Figure 20; Mortimore and Wood, 1986), where it is taken to mark the base of the Upper Chalk.

The Bridgewick Marls define the lower limit of a succession of nodular and incipient hardground chalks, the Kingston Beds, named after the track-cutting from Swanborough above Kingston near Lewes [387 075]. The Kingston Beds, 12.5 m thick at Bridgwick pit, yield sparse elements (including ammonites and gastropods) of the rich '*reussianum* fauna' of aragonite-shelled molluscs that characterises the topmost hardground of the Chalk Rock in the Chilterns and adjacent areas (see below). The equivalent beds at Dover are highly fossiliferous by contrast, but the sequence there is relatively condensed and incomplete compared with that in the Lewes area. The upper Kingston nodular beds are characterised by remarkable subvertical digitate thalassinoidean burrow-form flints, 30 mm in diameter, which penetrate the nodular chalk sequence for a depth of up to 4 m. The distinctive feature of these flints is the fact that the burrow wall is replaced by white flint meal and that the silicification extends outside this layer. The net effect is to produce what appears to be an annulus of chalk within a solid flint; locally the central core is missing, and for some part of its length the flint is thus apparently tubular. These very distinctive flints can be traced at this level throughout the Anglo-Paris Basin, and provide a useful stratigraphical marker. The upper Kingston nodular beds, with their associated digitate 'tubular' flints, terminate upwards in a thick marl seam, the Lewes Marl, which contains abundant large *Micraster leskei*, and is rich in crinoid debris.

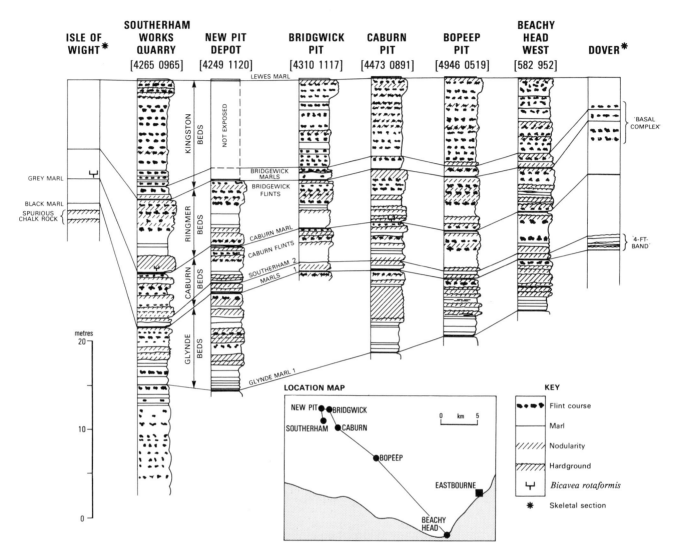

Figure 20   Correlation of some important sections in the lower part of the Lewes Nodular Chalk Member

The South Street Beds (Figure 12), 6.5 to 9 m thick, comprise a basal nodular bed succeeded by massive chalk with a weak griotte texture which is overlain by nodular hardgrounds with marl separations, the Lewes Nodular Chalks. Above these there is an abrupt change in the sediment type, with the entry of Bänderkreide (Cuilfail Bänderkreide) at the base of the Navigation Beds. A comparable change is observable at Dover, but the *Zoophycos* traces in the successions around Lewes are locally silicified, preserving details of the trace-fossil (cf. Bromley and Ekdale, 1984, fig. 9b): this is particularly well seen in the Southerham Works Quarry [4265 0965]. The *Zoophycos* chalks terminate in a sequence of closely spaced strongly mineralised hardgrounds, named the Navigation Hardgrounds after the former Navigation pit [425 100]. The Navigation Hardgrounds equate with the so-called 'Top Rock' of the Dover succession, and are overlain by a closely spaced pair of discrete marl seams, the Navigation Marls.

The succeeding Cliffe Beds (about 3 m thick), Hope Gap Beds (about 4 m), Beeding Beds (about 8 m), Light Point

Beds (about 5 m), and Beachy Head Sponge Beds (about 9 m) comprise dominantly nodular flinty chalks (Figure 21). Each of these subdivisions is typically capped by a mineralised hardground, of which the Cliffe Hardground is the most conspicuous in this district.

There is a significant increase in faunal diversity near the base of the Glynde Beds, with the entry of *Sternotaxis planus* and '*Micraster corbovis* of *lata* Zone type' (Stokes, 1975) in the lower Glynde nodular chalks, and the entry of the coccolith *Eiffelithus eximius* in the Glynde Marl itself (Figure 19). *Inoceramus cuvierii* s.s. continues up from the underlying New Pit Beds into the griotte chalks between the main Glynde Marl and Glynde Marl 2, and then appears to cut out. The upper Glynde nodular chalks are locally very fossiliferous, yielding *S. planus*, *Gauthieria radiata* and a diverse inoceramid assemblage including *I. apicalis*, *I.* cf. *flaccidus* and a rounded form (*I. labiatus latus*), tentatively assigned to the group of *Mytiloides? fiegei*. These are associated with a sparse heteromorph ammonite assemblage, including species of *Scaphites* and *Sciponoceras*. The type of *Inoceramus lamarcki* was

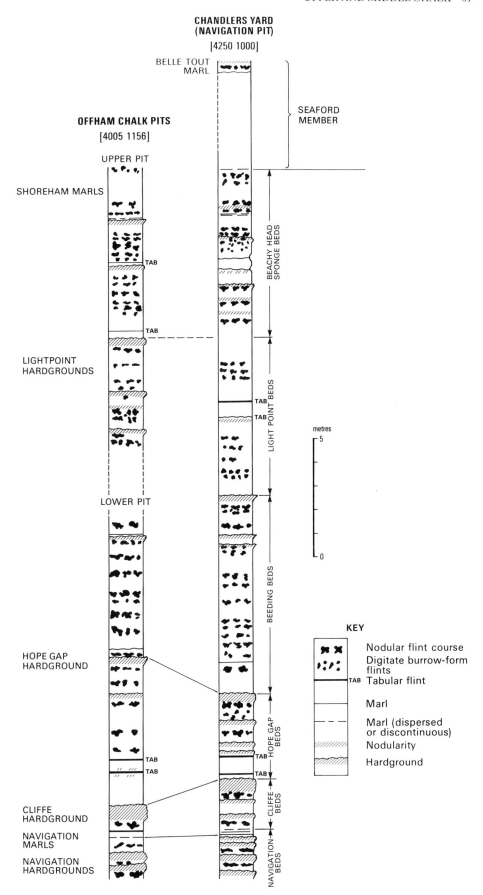

Figure 21 Sections in the upper part of the Lewes Nodular Chalk Member near Lewes

collected at Dover from the flints beneath the 4-ft-band, but *I. lamarcki* preserved in flint has not yet been noted at this horizon in the Lewes district.

The interval between the paired Southerham Marls at the base of the Caburn Beds marks the upward limit of the range of *Coscinophragma sp.*, immediately above its acme occurrence in the Southerham Marl 1 (see p. 59). This interval also yields *Gauthieria radiata* and small ammonites including *Lewesiceras sp.* and *Subprionocyclus sp.*, the latter genus suggesting that this horizon falls within the Upper Turonian *Subprionocyclus neptuni* Zone. A juvenile *Inoceramus inaequivalvis* was found at this level in the Caburn pit. The microsterid *Epiaster michelini* Agassiz *sensu* Stokes is known from this level at Dover (Stokes, 1975), but the lowest records of this species in the Lewes district are from the Caburn Sponge Beds immediately beneath the Caburn Marl at the top of the unit. Just as in the case of the upper Glynde nodular chalks, a limited aragonite-shelled fauna is also preserved in the nodular Caburn Sponge Beds. This fauna includes gastropods (*Bathrotomaria*) and small ammonites (*Yezoites* [*Otoscaphites*] and *Sciponoceras*) at the Bridgwick pit, and the figured specimen (BGS GSM117333) of *Romaniceras deverianum* from the very degraded exposure in the Firle pit (Wright and Kennedy, 1981, pl. 16, figs. 2a–b) is also believed to come from this horizon. The last species provides a link between the ammonite-poor chalk facies successions of southern England and the stratotype Turonian in the southern part of the Paris Basin (Touraine). The inoceramid fauna of the Caburn Beds is poorly known at present, but large forms close to *I. securiformis* occur at the level of the Caburn Sponge Beds.

In the Ringmer Beds the lower nodular beds are characterised by the occurrence of the small wheel-like bryozoan *Bicavea rotaformis*. The discovery of this species at both Caburn and Southerham Works Quarry during the present re-survey established the correlation between the succession in the Lewes district and the very much condensed succession in the Isle of Wight. It now appears that the Caburn Marl of Sussex equates with the 'Grey Marl' of the Isle of Wight with its overlying *Bicavea* Bed (Brydone, 1917). Arising from this correlation, the 'Black Marl' of the Isle of Wight equates with the lower of the paired Southerham Marls, and the underlying paired glauconitised hardgrounds known as the 'spurious Chalk Rock' (Rowe, 1908) are probably the condensed expression of the weakly indurated Glynde nodular chalks. The correlation between the Lewes, the Isle of Wight and the Dover successions is shown in Figure 20.

The Ringmer Beds are rich in echinoids, notably *Sternotaxis planus*, which occurs throughout, as do the brachiopods *Gibbithyris? sp.* and *Orbirhynchia dispansa*. The upper nodular chalks yield large forms of '*Epiaster michelini*' associated with '*Micraster corbovis* of *lata* Zone type', together with the crinoid *Isocrinus granosus* and large inoceramids of the *I. cuvierii* lineage. This level at Dover marks the entry of the nannofossil *Marthasterites furcatus*, a datum taken by many workers to recognise the base of the Coniacian Stage, but in southern England clearly falling well within the Upper Turonian (Bailey and others, 1984).

The upper limit of the range of *Terebratulina lata* is found in the Bridgewick Marl 1 at the base of the Kingston Beds. The interval between the paired Bridgewick Marls is characterised by relatively common *Gauthieria radiata* and *Sternotaxis planus*, together with an acme occurrence of *Orbirhynchia dispansa*. Micrasterids are scarce at this level. A significant increase in the faunal diversity is found in the lower Kingston nodular chalks. This is marked by the entry of the rhynchonellid brachiopod genus *Cretirhynchia*, and by the appearance of *Micraster leskei* d'Orbigny *non* Desmoulins (Figure 19). These beds yielded *Allocrioceras billinghursti* and *Yezoites* [*Otoscaphites*] *sp.* at the Bridgwick pit.

The interval between the lower and the upper Kingston nodular chalks contains several bands of *S. planus*. The three closely spaced bands of nodular chalk comprising the upper Kingston nodular chalks are characterised by '*Cretirhynchia*' *minor*, associated with *C. cuneiformis*, *Orbirhynchia reedensis* and *Gibbithyris spp.* These brachiopods are particularly common in the lowest of the three nodular beds, associated with *Mytiloides fiegei* [? = *M. incertus*], *Lewesiceras mantelli* and *Scaphites geinitzii*. In the higher two nodular beds, the inoceramid assemblage comprises inflated *Mytiloides sp.*, provisionally attributed to *M. carpathicus* s.l. and subordinate *M. fiegei*. There is a marked concentration of inoceramids above the second nodular bed, and a second one above the top nodular bed immediately beneath the Lewes Marl. The highest nodular bed has yielded a specimen of *Subprionocyclus sp.*

The fauna of the upper Kingston nodular chalks correlates in part with the so-called '*reussianum* fauna' of the pebble bed of the Hitch Wood Hardground (Bromley and Gale, 1982) at the top of the Chalk Rock hardground sequence in the Chilterns and adjacent areas. The Chalk Rock '*reussianum* fauna' is characterised by an abundant and diverse assemblage of originally aragonite-shelled molluscs (notably ammonites and gastropods), preserved as moulds, together with sponges, corals, brachiopods, bivalves (both calcitic and aragonitic) and echinoids. It is assigned to the late Turonian *Subprionocyclus neptuni* ammonite zone. This '*reussianum* fauna' is comparatively well developed in the Kingston nodular chalks of the Dover succession, but in the Lewes district it is only weakly and locally represented, being restricted to poorly preserved *Bathrotomaria perspectiva*, trochid gastropods and few ammonites.

The Lewes Marl is rich in crinoids (*Bourgueticrinus*) and contains abundant *Micraster leskei* (the 'large form' of Stokes, 1975). In the beds above, *M. leskei* passes into forms transitional between *M. leskei* and *M. normanniae* (Stokes, 1975), attributed by Drummond (1985) to *M. precursor*.[1]

Sternotaxis planus ranges up to the top of the South Street Beds, and there are sparse occurrences of *Echinocorys sp.* throughout. '*Cretirhynchia*' *minor* ranges up from the

---

1 One school of thought believes that the many established 'species' of *Micraster* are little more than ecomorphs of a single long-ranging species. Such an approach is contrary to that of Drummond (1985), who recognised a succession of species within an evolutionary lineage, and thereby attempted to portray in updated form the classic sequence of changes described by Rowe (1899). Drummond's splitting into species and subspecies is not acceptable to many other workers, and it should be noted that Rowe himself recommended that his form-name '*praecursor*' should not be regarded as having specific status. In view of this controversy, a relatively conservative position is adopted here, close to that of Stokes (1975), but which differs in that Stokes's *M. decipiens* is treated as a synonym of the earlier name *M. cortestudinarium* (see footnote, p. 63).

Kingston Beds into the lower part of the South Street Beds, and is represented by comparatively large individuals. The latter beds are particularly characterised by large flat *Mytiloides sp.* that provide a marked contrast with the more inflated forms of the underlying Kingston Beds. These elongated inoceramids are strikingly similar to *Mytiloides mytiloides* from the early Turonian Holywell Beds and, in the absence of stratigraphical data to the contrary, could easily be confused with them. This South Street Beds *Mytiloides* assemblage is found in late Turonian successions elsewhere, notably in the USA and in Lower Saxony, GFR, and in all areas replaces inoceramid assemblages dominated by *I. costellatus* and *M. fiegei* such as characterise the Kingston nodular chalks and correlatives. Aragonitic faunas are poorly represented in the South Street Beds: *Lewesiceras sp.* was collected from just above the Lewes Marl, and the Lewes Nodular Chalks have yielded *Bathrotomaria sp.*, an indeterminate nautiloid and *Lewesiceras sp.*

Throughout the district, there is an abrupt and conspicuous change in the style of bioturbation above the Lewes Nodular Chalks. The South Street and underlying beds are characterised by chalks with the trace-fossil (ichnogenus) *Thalassinoides*. In the Navigation Beds, *Thalassinoides* is still present, but the burrows have been secondarily burrowed by *Zoophycos*, implying two distinct phases of sedimentation. The appearance of *Zoophycos* at this level in the succession produces an approximation to the 'Bänderkreide' ichnofacies of the Maastrichtian of northern Germany, and provides a lithostratigraphical marker that can be traced throughout the northern part of the Anglo-Paris Basin. Concomitant with, and perhaps in response to the change in depositional environment implied by the presence of *Zoophycos*, there is a marked decrease in macrofaunal diversity, and a significant change in the echinoids. The smaller and more inflated long-ranging *Sternotaxis planus* found in the underlying beds is replaced by the large flat-based species *S. placenta*, which occurs abundantly in and characterises the Cuilfail *Zoophycos* chalk. There is also a significant change in *Micraster*, with the replacement of *M. 'precursor'* by *M. normanniae*.

The Navigation Hardgrounds are rich in *Echinocorys gravesi*, associated with an apically pointed variant of this species that seems to be restricted to this level. The equivalent of the highest of the Navigation Hardgrounds at Dover yielded a small ammonite fortuitously preserved inside a broken echinoid test. This ammonite has been tentatively identified as *Forresteria (Harleites) petrocoriensis*, the index fossil of the lowest of the three currently accepted Coniacian ammonite zones (see Gale and Woodroof, 1981; Bailey and others 1983; 1984). No ammonites have been found in the Navigation Hardgrounds of the Lewes district.

It should be noted that the boundary taken by Rowe (1900) between the *planus* and *cortestudinarium* zones in the Dover cliff sections approximates to the base of the Navigation Beds. Subsequent workers (e.g. White, 1928; Stokes, 1975; Rawson and others, 1978) have, however, drawn the boundary at a level equivalent to the top of the Navigation Hardgrounds as defined by the 'Top Rock' on Langdon Stairs near Dover. The base of the Navigation Beds also corresponds to the base of the Senonian as currently interpreted in France, and thus to the base of its lowest constituent stage,

the Coniacian. Although the uppermost Navigation Hardground must be assigned to the basal Coniacian on the evidence of the single *Forresteria* from Dover, the attribution of the underlying part of the Navigation Beds succession is controversial, and inoceramid (and ammonite) evidence from correlative beds elsewhere in Europe tends to support a late Turonian age (see discussion by Wood and others, 1984).

The remainder of the Lewes Chalk Member above the Navigation Beds can be divided biostratigraphically into two parts. The interval between the Navigation and Hope Gap hardgrounds is characterised by the small smooth weakly plicate rhynchonellid '*Cretirhynchia*' *subplicata*, together with *Micraster* transitional between *M. normanniae* and *M. cortestudinarium*[1]. This interval also yields small geniculate inoceramids near *Cremnoceramus inconstans* [Woods, 1912, vol. II, fig. 43] and *C.? waltersdorfensis*, for example at the Offham small pit. *C.? waltersdorfensis* and *C.?* cf. *rotundatus* are known from the equivalent of the Cliffe Hardground at Dover and Shoreham cement works in the Brighton district, and indicate a level low in the Coniacian.

At the Hope Gap Hardground there is a significant faunal change. The Hope Gap Hardground marks the lowest occurrence of *Micraster gibbus* [= *M. (Isomicraster) senonensis*] and the overlying coarse detrital chalks are rich in the thintested echinoid *Cardiaster cotteauanus*. Above the Hope Gap Hardground, *Micraster* transitional between *M. normanniae* and *M. cortestudinarium* is replaced by *M. cortestudinarium*. In the highest beds of the member, above the upper Light Point Hardgrounds, *M. cortestudinarium* is stated by Drummond (1985) to pass via a series of transitional forms into *M. turonensis*, which is first encountered above the Shoreham Marls at the base of the traditional *M. coranguinum* Zone.

The Hope Gap Hardground also marks a significant change in the inoceramid assemblages, with the entry of large *Cremnoceramus* of the *C. erectus – C. deformis – C. schloenbachi* lineage. There is only limited information regarding the stratigraphical distribution of the individual species comprising this lineage, but 'late forms' (*sensu* Kauffman) of *C. erectus* are known from immediately above the Beeding Hardground, and large forms close to *C. schloenbachi* occur in the Beachy Head Sponge Beds.

The *Trochiliopora* Bed, attributed by Gaster (1920) to a low position in the *coranguinum* Zone, is probably situated at about the level of the Beachy Head Sponge Beds (i.e. near the top of the *cortestudinarium* Zone as conventionally interpreted), to judge from Gaster's reference (1937a) to yellow nodular chalks and inoceramid shell debris. This bed, stated to be 10 feet [3 m] thick, contains many bryozoa, notably *Trochiliopora gasteri*. It was not recognised in the present investigation.

1 The only extant specimen from Goldfuss's original syntype material of *M. cortestudinarium* is morphologically indistinguishable from the later-named *M. decipiens*, and cannot have come from any of the localities cited by Goldfuss, viz. Maastricht, Coesfeld and Quedlinburg. Although the specimen is free from matrix, the interior of the test contains a recrystallised chalk suggesting provenance either from the Anglo-Paris Basin (i.e. the same as that of *M. decipiens*) or from the correlative chalk facies of Lüneburg in northern Germany (Wood and others, 1984). In view of the almost certain conspecificity of *M. cortestudinarium* and *M. decipiens*, there is no need to replace the long-established *M. cortestudinarium* Zone by that of *M. decipiens* as used by Stokes (1975) and French workers.

SEAFORD SOFT CHALK MEMBER

The Seaford Member comprises generally homogeneous chalks, with frequent courses of large nodular flints, some of which are almost continuous and provide stratigraphical marker horizons; sponge-beds marking greater or lesser breaks in sedimentation are found throughout the succession. The base of the member is taken above the upper of the paired Shoreham Marls, and the top is marked by the surface beneath the lowest of the Buckle Marls (p. 66). The Bänderkreide texture of the underlying chalks is replaced by strong burrow-mottling.

The only complete sequence of the Seaford Member is found at Seaford Head in the Eastbourne district, where it is about 80 m thick, but incomplete sections are exposed in the cliffs of the Seven Sisters. Within the Lewes district, exposures are seen in the Southerham Works Quarry, the Tarring Neville quarry complex [446 034], the Juggs Lane road cutting [401 093] and a small roadside section near Piddinghoe [4325 0323] (Figure 22).

Compared with the greater part of the member, which is essentially devoid of marl, the basal beds (the Belle Tout Beds) are relatively marly, and include the Belle Tout Marls, named after a locality on the coast near Birling Gap in the Eastbourne district. The Belle Tout Beds are about 20 m thick at Lewes. This part of the succession terminates in a conspicuous semi-continuous tabulate nodular flint, the Seven Sisters Flint Band, which lies at the base of the Cuckmere Beds and is the correlative of the East Cliff Semi-tabular Flint of the equivalent Dover sequence (Gale and Smith, 1982). The Seven Sisters Flint forms a conspicuous marker in the highest beds of the Southerham Works Quarry (p. 71). It was this flint that Barrois (1876) mistook for the similar Whitaker's 3-inch Tabular Flint Band, which is found in the Thanet coast section at a somewhat higher stratigraphical level within the Seaford Member. For this reason, Rowe (1900, p.330) referred to the tabular flint of the Sussex coast as the 'spurious tabular'. In the beds above the Seven Sisters Flint Band, two distinctive markers occur (Figure 22), the Tarring Neville Flint and the Cuckmere Sponge Bed. The latter is a glauconitised and phosphatised hardground and was mistakenly correlated by Barrois (1876) with a sponge-bed, subsequently named the 'Barrois Sponge Bed', some 6 m above the Whitaker's 3-inch Tabular Flint on the Thanet coast: Rowe (1900) therefore called this bed the 'spurious (Barrois) sponge-bed'. In the Lewes district, the Cuckmere Sponge Bed was formerly exposed in the Juggs Lane road cutting.

The Michel Dean Flint Band marks the base of the Haven Brow Beds. This is the first of seven conspicuous flint bands, collectively known as the Seaford Flints. This flint and succeeding courses (Baily's Hill, Flat Hill) are each overlain by or associated with a concentration of inoceramid shells (Figure 22 and see below, p. 73). These three flint courses, equate in ascending order with the flint below the lower *Cladoceramus* band, the Chartham Flint and the Bedwell Columnar Flint of the Thanet coast succession (see Bailey and others, 1983, fig. 2; 1984, fig. 2). The columnar character of the Flat Hill Flint is well seen in the Lewes district at the top of the Tarring Neville section. These columnar flints are comparable with the paramoudra flints found much higher

in the Chalk, notably in Norfolk, but the central *Bathichnus paramoudrae* burrow, glauconitised in the Norfolk examples, is weakly mineralised in the columnar flints of southern England and consequently is barely detectable. In the Juggs Lane section, three weakly glauconitised sponge-beds correspond to the inoceramid concentrations associated in the coast sections with the three flint bands; there is no indication of these sponge-beds in the Tarring Neville section.

The Haven Brow Beds are poorly exposed in this district but at Seaford, where they are 24 m thick, these flinty chalks expose the full sequence of Seaford Flints and several discontinuous sponge beds.

The Seaford Member is co-extensive with the traditional *Micraster coranguinum* Zone. The zonal index does not appear until about the level of the Seven Sisters Flint Band, the underlying Belle Tout Beds being characterised by (*fide* Drummond) *M. turonensis*, including forms transitional to *M. coranguinum*. In terms of the old classification, the appearance of *M. coranguinum* approximately corresponds to the boundary between the 'lower third' and the 'upper two-thirds' of the zone. The zonal index occurs in bands throughout the Cuckmere and Haven Brow beds, passing through the progressive morphological changes described by Rowe (1899). *M. coranguinum* is, however, rather scarce in the Seaford Member of the Lewes district, and the inoceramid sequence proves to be of much greater use in zonal subdivision.

The Belle Tout Beds are characterised by *Volviceramus* and *Platyceramus*. In the equivalent succession in Kent, the lowest member of the *Volviceramus* lineage, *V. koeneni*, enters together with *Platyceramus mantelli* in a coarse shell-detritus-rich bed a short distance above the equivalent of the higher of the paired Shoreham Marls. The underlying chalk is characterised by small thin-shelled inoceramids which are possibly attributable to *Cremnoceramus*. In the Lewes district, the first *Volviceramus*, represented by *V.* ex gr. *involutus*, and *Platyceramus* occur at the level of the Belle Tout Marls. *Volviceramus* is found throughout the Belle Tout Marls–Seven Sisters Flint Band interval, but tends to occur at restricted horizons (Figure 22). The main concentration is immediately above the Seven Sisters Flint Band, and there is a second concentration some 1 to 2 m beneath the flint. Observations in other areas (Seaford Head and Dover) indicate that there is a significant reduction in both size and abundance of *Volviceramus* above the Seven Sisters Flint Band, following its size maximum in the belt beneath the flint. The *Volviceramus* at this level possess anteroposteriorly elongate right valves, in marked contrast to the nearly circular right valves found in *V. involutus* s.s.

Volviceramus ranges up into the Cuckmere Beds to an horizon between the Cuckmere Sponge Bed and the Michel Dean Flint. *V. involutus* s.s. has not been noted in the Lewes or adjacent districts, and in fact the stratigraphical provenance of the few specimens known from other areas is uncertain. Immediately overlying the Michel Dean Flint is a concentration of large thick-shelled *Cladoceramus sp.* with a distinctive pinkish purple colouration to the shell. A minor concentration of what is assumed to be the same species was observed in the Juggs Lane section above the Baily's Hill Flint, but is apparently absent elsewhere.

In the Lewes district, these two inoceramid horizons are associated with the lowest two of the three Juggs Lane

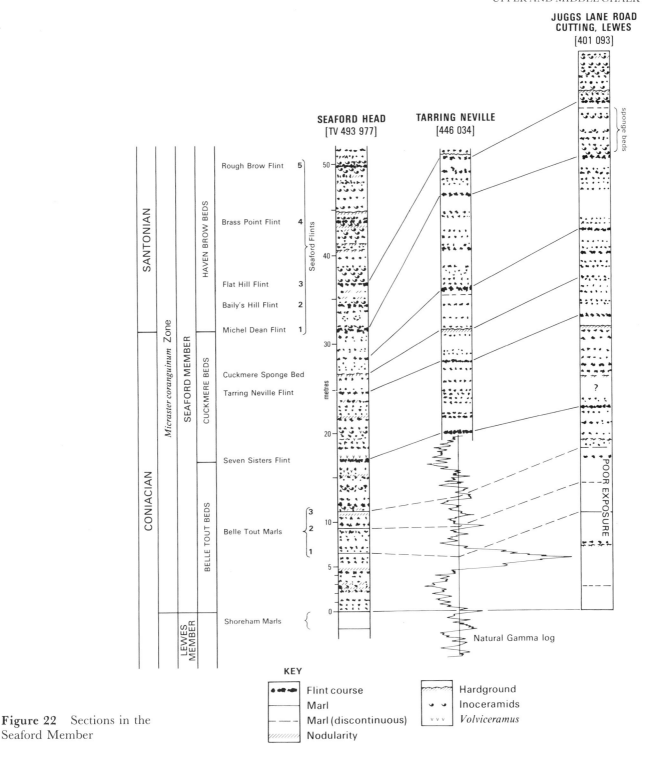

Figure 22 Sections in the
Seaford Member

Sponge Beds. There is a marked increase in faunal diversity at the Baily's Hill/Chartham Flint. In Kent, this is marked by the entry of the foraminifer *Stensioeina granulata polonica* and the macrofossils *Gibbithyris ellipsoidalis, Cordiceramus cordiformis, Spinaptychus* and *Conulus albogalerus* (see Bailey and others, 1983, p.37). *Conulus albogalerus* is an important component of the higher part of the *coranguinum* Zone fauna above this level. At the corresponding horizon in the Juggs Lane section, the inoceramid-rich sponge beds yielded *G.*

ellipsoidalis and *Orbirhynchia pisiformis*, together with *Bourgueticrinus granulosus*, radioles of *Tylocidaris clavigera*, the large thin-tested echinoid *Cardiotaxis aequituberculatus* and thin-tested *Echincorys sp.*

In the Juggs Lane section, thin-shelled *Cladoceramus undulatoplicatus* enters in strength in the uppermost of the three sponge beds immediately overlying the Flat Hill Flint. The same *Cladoceramus* flood is seen in the Tarring Neville section, and in the coast section at Seaford Head, as well as at the base of the Bedwell Columnar Flint Band on the Thanet coast. A magnificent example of *Tylocidaris clavigera*, with dental apparatus and radioles, was collected from this horizon at Tarring Neville. The two *Cladoceramus* bands noted above are important marker horizons that can be traced throughout southern England and the northern part of the Paris Basin. The base of the Santonian Stage, in chalk facies, is provisionally recognised at the lower of the two *Cladoceramus* bands (see discussion by Bailey and others, 1984).

The upward limit of the range of *C. undulatoplicatus* is not known, but it has not so far been found above the Brass Point Flint (see Figure 22). Above this level occur large thick-shelled inoceramids. In the succession from the Flat Hill Flint to just above the Rough Brow Flint, the equivalent of Whitaker's 3-inch Flint Band of Thanet, occur four well defined and laterally continuous *Conulus* horizons with small forms of *C. albogalerus*. The highest Chalk exposed in the Lewes district, at Piddinghoe [435 030], contains two further *Conulus* bands, characterised by large acutely pyramidal variants of *C. albogalerus*. This occurrence compares with the *Conulus* Bed of the Thanet coast, situated some 4 m beneath the entry of *Uintacrinus socialis* at the base of the *U. socialis* Zone. The biostratigraphy of the *socialis*, *Marsupites testudinarius* and succeeding zones will not be considered further here, but will be described in the memoir for the adjacent Brighton district (in preparation).

NEWHAVEN GRIOTTE MARL CHALK MEMBER

The base of the Newhaven Member is drawn at Splash Point, Seaford Head [4899 9805], at the lowest marl seam of a group of several paired marls termed the Buckle Marls. Above this a remarkable change in lithology takes place. In contrast to the featureless, soft and medium-hard chalks with regular flint courses comprising the Seaford Member, the Newhaven Member is characterised by griotte chalks with abundant marl seams, extensive flint-free bands and hard chalks. Many of the marl seams form closely spaced pairs. Although the member is relatively flinty, large conspicuous semi-continuous flints such as those characterising the Seaford Member are missing. But for the presence of flint, the Newhaven Chalk represents a return to lithologies which seem to parallel to some extent those of the Ranscombe griotte chalks. The lowest 20 m of beds of the Newhaven Member are present in drift-free ground around Telscombe [404 033] and Piddinghoe [430 030] but there are no exposures.

CJW,RNM,RDL

THICKNESS VARIATIONS WITHIN THE CHALK

Recent work (Mortimore, 1979; 1986a) has demonstrated that considerable variations in thickness occur within parts of the Chalk succession in the Lewes district. The Chalk Marl, the Lewes Member, and the lower part of the Seaford Member show significant northward increases in thickness from the coastal sections of Beachy Head and Seaford into this district. Near Lewes a reversed trend is observed in some members (e.g. in the Lewes Member, Figure 23). Other parts of the sequence, for example the Grey Chalk, appear to retain an almost constant thickness.

Where thinning occurs, it is marked by increased erosion, non-sequences and the presence of mature hardgrounds. The areas of maximum thinning and the development of hardgrounds appear to be near Eastbourne, and to the north-west of Brighton. Consequently it is thought that two swells, near Brighton and Eastbourne, with an intervening trough to the south of Lewes, existed during much of the late Cretaceous (Mortimore, 1979). The swells may reflect fault-bounded Hercynian structures in the basement (Lake, 1975b), which were periodically rejuvenated during the deposition of the Chalk. The trough to the south of Lewes appears to have been a repository for material swept as slurries from the adjacent swells, and this may explain the relatively common occurrence of lignitic and vertebrate material in the Chalk of this area.

The difficulty in defining the top of the Chalk Marl has been discussed above (p. 52). Because of this it is not yet possible to standardise lithological and geophysical borehole logs in this subdivision. It is apparent in exposures, however, that the Chalk Marl varies in thickness from 22 m at Beachy Head to over 40 m at Rodmell [440 064]; about 42 m are present in the Southerham area. Details of condensed sequences and thickness changes within the Chalk Marl have been presented previously (pp. 50–52).

The Ranscombe Member is about 80 m thick in the Rathfinny water-wells [4948 0170; 4971 0155] and about 90 m at Beachy Head [5881 9545], but possibly as little as 60 m thick near West Firle [470 071]. A condensed sequence with phosphatic chalks and mature hardgrounds has been recognised in the Mount Caburn area, where these beds are about 65 m thick.

The thickness and lithological variations of the Lewes Member are shown in Figure 23. In this case an additional secondary trough is indicated near Mount Caburn, possibly coaxial with the Kingston–Beddingham Anticline.

The Cuckmere Beds of the Seaford Member show a northward thickening from the coast at Seaford Head [493 977] (19.5 m) to Tarring Neville [446 034] and Piddinghoe [4325 0323] (31 m). In the Juggs Lane cutting [401 093] the presence of small-scale slumped beds in the Cuckmere Beds suggests a trough-margin situation. RNM,RDL

DETAILS

Lower Chalk

GLAUCONITIC MARL

Few noteworthy exposures of the Glauconitic Marl remain, though White (1926) and Gaster (1951) recorded former sections of historical interest. Lithologies typical of those described above (p. 49) were encountered in the BGS boreholes at Glyndebourne [4420 1141], 1.26 m thick; at West Firle [4676 0809], 1.24 m; and at Alciston [5045 0553], 1.74 m.

The best exposure, and the only one to show the junction with the Gault, was seen in the southern part of the Rodmell (Asham) Cement Works clay-pit [4413 0711]. The section recorded in 1973 was: Gault (see p. 42), overlain by hard glauconitic marlstone 0.3 m, and friable glauconitic marl, ochreous and green 0.9 m (see also p. 49). A small section [4604 1129] about 100 m north of Oldhouse Farm showed about 0.5 m of Glauconitic Marl.

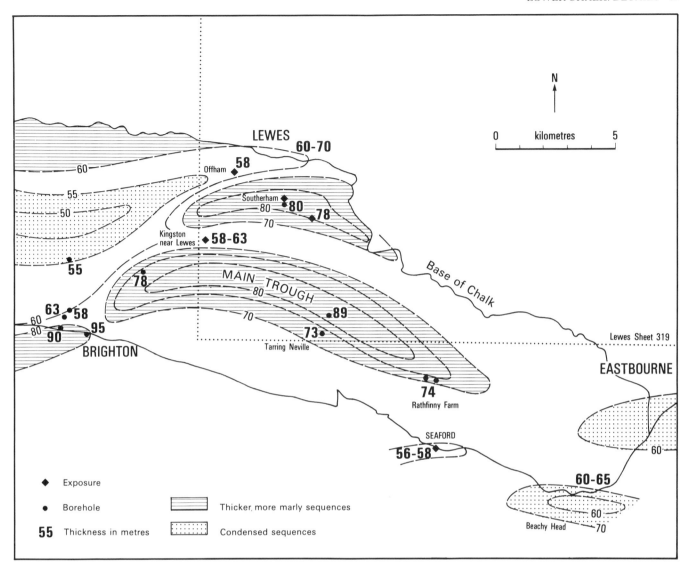

Figure 23 Isopachytes and lithological variations of the Lewes Nodular Chalk Member

Gaster's (1951) records of Glauconitic Marl near Preston Court [4590 0760] and West Firle village [4661 0736] are apparently erroneous. Excavations [4703 0811] alongside the new road approximately 50 m NE of Gibraltar Farm, West Firle, revealed up to 1.5 m of dark green, slightly sandy Glauconitic Marl. Up to 0.9 m of weathered Glauconitic Marl was exposed in the overgrown north bank of the lane [5051 0566] 120 m NW of Alciston church.   BY,RDL

CHALK MARL AND GREY CHALK

About 6 m of thickly bedded off-white chalk were seen in the old pit [4114 1111] 150 m NW of South Malling church and there were a number of exposures of chalk at roughly the same stratigraphical level in the overgrown sides of the disused railway cutting [4135 1109] north and north-west of the church. White (1926) assigned the chalk in this pit to the *Holaster subglobosus* Zone, though Gaster (1951) considered it to be part of the *Inoceramus labiatus* Zone of the Middle Chalk. The recent revision survey showed the strata to be stratigraphically below the Melbourn Rock. There is a section, 4.6 m high, of thickly bedded off-white chalk with a number of thin

marly beds up to 0.2 m thick in the steep river cliff [4160 1138] 480 m NE of South Malling church.

The Glyndebourne Borehole [4420 1141] penetrated 46.36 m of Lower Chalk beneath 2 m of Head. All but the top 7.55 m of Chalk was cored (Figure 15). Between 27.72 and 47.10 m the succession is markedly rhythmic as described above (p. 50). The rhythms range between about 0.2 and 1.4 m in thickness, with the thinner and better developed rhythms being found below 35.00 m. Individual hard limestone beds rarely exceed 0.2 m in thickness and usually constitute about one third of the rhythm. Most, but not all, of the tops of the hard beds display sharp contacts with the overlying sediment, but with the exception of one bed at 45.02 m they appear to pass down into the underlying sediment without any perceptible break. This contrasts with the observation of Destombes and Shephard-Thorn (1971) who, in their study of the Chalk near Dover, described the harder limestone beds at this level as having sharp bases and tops.

Some of the most extensive sections of Lower Chalk exposed in the district may be seen in two large quarries at Southerham (Figure 16). Some confusion has arisen over the naming of these pits. The older [4320 0905], on the north side of the Lewes to Eastbourne

(A27) road, was disused when White described it in 1926, and was known then as the Southerham Grey Pit. This pit has since been re-opened and it is now known as Machine Bottom Pit; both it and the newer pit [4280 0900], between the road and the railway, have been greatly enlarged, the newer, more westerly pit being named 'Southerham Grey Pit' on the New Series six-inch map.

In the Southerham Grey Pit [4280 0900] a total of some 50 m of Lower Chalk is exposed (Figure 16; Plate 6). Kennedy (1969, fig. 10, pp.497, 499) has given a detailed palaeontological description of this sequence; his bed nomenclature is shown in Figure 16 so far as correlations can be established. Subsequent work has further refin-ed Kennedy's interpretation. The Chalk Marl is highly fossiliferous and the limestones, in particular, yield well-preserved uncrushed ammonites and inoceramids. The zonal and subzonal boundaries are, however, difficult to determine. The occurrence of *Inoceramus* ex gr. *virgatus*, both in a band in flood abundance and in the limestone above (Figure 16), suggests a correlation with the upper part of the *M. saxbii* Subzone as recognised at Folkestone. There is no positive evidence of the *M. dixoni* Subzone, but poorly preserved small *Mantelliceras* occur with *Schloenbachia* at the base of the overly-ing marl and in the 2 m of beds above, including a thin limestone. This marl also marks the upper limit of the range of *M.? rectifrons*.

The next limestone in the succession has yielded a rich and well preserved ammonite fauna, comprising predominantly *Schloen-bachia*, with sparse *Acompsoceras*, *Hyphoplites* and *Turrilites* cf. *borssumensis*. The presence of *Hyphoplites* indicates that this bed is of Lower Cenomanian age, despite the apparent absence of *Mantelliceras*. The rich inoceramid assemblage comprises '*Inoceramus*' ex gr. *crippsi*, including '*I.*' aff. *crippsi* and '*I.*' *crippsi hoppenstedtensis*, with subordinate *I. virgatus* and *I.* cf. *conicus*. In con-trast to this bed, the marls above are almost devoid of macrofossils, apart from poorly preserved moulds of small *Schloenbachia* and fragments of inoceramids.

The entry of common *Turrilites scheuchzerianus* in the next promi-nent limestone (Figure 16) indicates the base of the Middle Cenomanian *A. rhotomagense* Zone (*T. costatus* Subzone), (cf. Ken-nedy, 1969, fig.10), although *Acanthoceras* itself first appears some 7 m higher in the succession, in the lower *Orbirhynchia mantelliana* band. This latter band is the lowest of the three Middle Cenoma-nian brachiopod bands, all of which are readily recognisable here. By analogy with the Rodmell succession ('Beddingham' of Carter and Hart, 1977), the top of the upper *Orbirhynchia mantelliana* band marks the mid-Cenomanian non-sequence, as defined by the entry of planktonic foraminifera at the base of Zone 11ii. About 1 m above this band, the boundary between the Chalk Marl and the Grey Chalk is abrupt (Plate 6), in contrast to other localities. Large *Acanthoceras* ex gr. *jukesbrownei*, indicative of the *jukesbrownei* Zone, are found above this boundary (Kennedy, 1969, fig.10), and so the interval between the highest brachiopod band and the base of the Grey Chalk lies in the *Turrilites acutus* Subzone.

A remarkable feature of this section is the presence of a shallow channel, with a lateral extent of some 100 m, that cuts down from a level 1.5 above the base of the Grey Chalk through 6 m of Chalk Marl (Figure 16) to the base of the band of diverse small brachiopods. This structure is inaccessible in the quarry face, but fallen blocks of the channel-fill show that the basal bed varies be-tween 10 and 70 cm in thickness, and that it fills *Thalassinoides* bur-rows in the underlying chalk. This basal bed comprises cemented coarse calcarenites with a high content of echinoid debris, and in-cludes angular chalk clasts, glauconitised and phosphatised pebbles, and much pelletal phosphate with small fish teeth. The trace-fossil *Teichichnus* also occurs. Reworked phosphatised *Orbirhynchia mantelliana*, preserved both as shells and as steinkerns, indicate derivation from the erosion of the upper *Orbirhynchia mantelliana* band.

In the lower part of the Grey Chalk there are conspicuous marl seams which preliminary work suggests may be used for regional correlation. This succession is characterised by large *Acanthoceras* ex gr. *jukesbrownei*, and there are several levels at which '*Inoceramus*' *atlanticus* occurs commonly. A massive unit, some 6 m thick, with scour structures, is found near the top of the Southerham Grey Pit and corresponds to the 'Jukes-Browne Bed 7' of the Dover–Folke-stone sequence; the lower limit of this unit at Southerham might be taken either at a marl 11 m above the base of the Grey Chalk, or at a second marl some 80 cm higher; detailed correlation, using all the marl seams in this part of the succession, points to the higher horizon as being the correct correlative. At Southerham, scour structures of the type that characterise 'Jukes-Browne Bed 7' are found at several levels in the underlying beds (see Figure 16).

The Machine Bottom Pit [4320 0905] exposes approximately 30 m of Lower Chalk and has been described by Barrois (1876), Jukes-Browne and Hill (1903), White (1926), Gaster (1951), and Kennedy (1969). The beds exposed here range from the upper part of the *Acanthoceras rhotomagense* Zone to the *Mytiloides labiatus* Zone.

The section exposed in 1973 may be summarised thus:

	Thickness m
MIDDLE CHALK (Ranscombe Griotte Chalk Member)	
[1]Melbourn Rock: Creamish white, hard, nodular chalk with marl partings; some of the nodular beds are markedly iron-stained	0.9
LOWER CHALK	
[1]Plenus Marls: Pale grey, very marly chalk with slight greenish tinge; two well marked hard beds up to 0.3 m thick; sharp erosional base	4.8 to 5.7
Grey Chalk: Greyish white chalk, massive, thickly bedded, with thin marly partings	29.2
Chalk Marl: Rhythms up to 0.9 m thick, consisting of medium grey marly chalk grading up into pale fawn hard chalk commonly with large uncrushed ammonites; the hard beds generally have sharp tops, commonly burrowed; scattered pyrite nodules	9.1

1 These beds, exposed high up in the north-east face of the pit, are inaccessible.

Individual rhythms in the Chalk Marl are, in places, less clearly defined in the Machine Bottom Pit than in Southerham Grey Pit, with certain hard beds having no clearly defined sharp top. As in the Grey Pit, the hard beds are locally very fossiliferous, a feature of the Machine Bottom Pit being the occurrence of large specimens of *Acanthoceras sp.*, several very large examples of which were seen in loose blocks.

The Grey Chalk at the Machine Bottom Pit is generally very similar to that of the Southerham Grey Pit (see Figure 16), though the section is less weathered and the individual beds are consequent-ly less well defined. Kennedy (1969) recorded an *Acanthoceras jukesbrownei* zonal assemblage from the lowest 10m, including numerous *A.* ex gr. *jukesbrownei*, *A. sp.*, *Protacanthoceras sp.*, *Schloen-bachia sp.*, and the nautiloids *Cymatoceras deslongchampsianum* and *Eutrephoceras expansum*. An occurrence of *Inoceramus pictus* overlying 'Jukes-Browne Bed 7' is noteworthy. The whole of the Grey Chalk succession above this is attributable to the *C. guerangeri* Zone. In the past, the Grey Chalk at Southerham was an important source of vertebrate remains. The top of the Grey Chalk is overlain by the Plenus Marls with a markedly discordant base.

Excavations [4275 0915] for the Lewes by-pass between the Southerham Grey Pit and the Machine Bottom Pit exposed Grey Chalk similar to that seen in the upper part of the Southerham Grey Pit. One bed of 'gritty' chalk here yielded abundant *Acanthoceras jukesbrownei*, together with other species of *Acanthoceras*; this bed ap-pears to be at a level between the base of the Grey Chalk and the base of the equivalent of 'Jukes-Browne Bed 7'.

The chalk-pits at Rodmell (Asham) Cement Works have also

Plate 6   Chalk Marl overlain by
Grey Chalk in the Southerham
Grey Pit (R. N. Mortimore)

been described by Kennedy (1969), but under the name 'Bed-dingham'. Alternating grey limestones and marls (Chalk Marl) overlain by massive grey chalk (Grey Chalk) were formerly exposed discontinuously in the northern pit [438 067], where Kennedy (1969, pp.502–503) recorded the presence of beds rich in *Orbirhynchia mantelliana* and *Sciponoceras* (the upper *Orbirhynchia mantelliana* band of this memoir) 3.5 to 5 m below the massive chalk. In the large abandoned pit [440 064] extensive sections exposed, in 1973, beds ranging from the Chalk Marl to immediately above the Melbourn Rock (Figure 17), though the higher parts of the sequence were generally inaccessible. Kennedy (1969, pp.538–589) listed the Middle Cenomanian macrofaunas from this locality, and Carter and Hart (1977, fig. 26 and p.83) provided microfaunal data. They showed that the boundary between the benthonic foraminiferal zones 11i and 11ii, corresponding to the mid-Cenomanian non-sequence, occurred at the top of the upper *Orbirhynchia mantelliana* band, about 3.5 m beneath the onset of massive

chalk at the base of the Grey Chalk. Kennedy attributed this upper-most part of the Chalk Marl above the upper *O. mantelliana* band to the *Turrilites acutus* Subzone of the *Acanthoceras rhotomagense* Zone, and noted that large ammonites from hard bands a few metres above the base of the Grey Chalk could be referred to *Acanthoceras* ex gr. *jukesbrownei*. Boreholes sunk in the floor of this pit proved difficult to correlate at about the level of the upper *Orbirhynchia mantelliana* band, and this problem may be related to channelling analogous to that observed in the Southerham Grey Pit (p. 68).

Newington's Pit [4605 0850], formerly known as Balcombe Pit, immediately south-east of Glynde Station, is the only quarry in Lower Chalk in the district that is still being worked. It provides continually varying exposures of beds between the upper *Orbirhynchia mantelliana* band and the equivalent of 'Jukes-Browne Bed 7'. The beds in the northern face of the pit dip at 3 to 5°NW and dips of around 5° slightly east of south were recorded in the south-east corner of the pit, suggesting that the axis of the Kingston–

Beddingham Anticline runs through the pit. Kennedy (1969, pp. 500–501) recorded a *Turrilites acutus* Subzone ammonite assemblage from the north side of the pit, and many representatives of the *Acanthoceras jukesbrownei* zonal assemblage — including the zonal index, *Austiniceras austeni*, common *Calycoceras spp.*, *Protacanthoceras sp.*, *Sciponoceras* aff. *baculoide*, *Schloenbachia sp.* and *Scaphites sp.* — from loose blocks estimated to come from 3 to 4 m higher in the sequence. Johnson (1901) recorded ten species of fish from this locality, including a group of 47 associated teeth of *Ptychodus decurrens* and a mandibular ramus of *Pachyrhizodus gardneri*.

The lowest 14.12 m of the Lower Chalk were cored in the West Firle Borehole [4676 0809], about 600 m SE of Newington's Pit. A typical Chalk Marl rhythmic sequence of hard limestones and softer marls (12.88 m) was proved overlying 1.24 m of Glauconitic Marl and the base of the Lower Chalk was penetrated at 18.12 m depth. The Chalk Marl was rich in ammonites, mainly *Schloenbachia sp.*, and *Monticlarella? rectifrons* occurred throughout. The lowest four hard beds in the sequence (below 15.47 m) ranged from 0.02 to 1.5 m in thickness and displayed sharp burrowed tops and bases, more nearly resembling the hard beds in the Chalk Marl described by Destombes and Shephard-Thorn (1971) than the equivalent beds seen in the Glyndebourne Borehole (p. 67). These hard beds contained abundant sponges and both they and the intervening more marly chalk were intensely burrowed. Above 15.47 m a well-marked rhythmic pattern was evident, closely resembling the equivalent part of the sequence at Glyndebourne, with hard beds generally passing down gradually to the underlying marly beds. The rhythms were up to about 1 m thick, with the top hard unit generally accounting for about one third of the thickness of each rhythm. Above 13.00 m sponges were not as abundant as at Glyndebourne. Many of the hard units exhibited sharp burrowed tops, though a number were seen to pass gradually upwards into the succeeding marly units.

During the construction of the new section of the Lewes to Eastbourne (A27) road at Gibraltar Farm [4690 0810] excavations showed a total of about 7.6 m of rhythmically bedded Lower Chalk, the hard beds yielding numerous well-preserved ammonites.

The lowest 14.32 m of the Lower Chalk was seen in the Alciston Borehole [5045 0553], 150 m WSW of Alciston church, which cored 12.58 m of Chalk Marl overlying 1.74 m of Glauconitic Marl. The Chalk Marl sequence of rhythmically bedded chalk was very like that in the West Firle Borehole, and contained *Monticlarella? rectifrons* throughout, indicating that only the lower part of the Lower Cenomanian was represented.

Up to 3 m of massive, off-white chalk, high in the Lower Chalk, were exposed in the high south bank of the lane [5213 0382] immediately south of Winton Farm, Alfriston.

The top 10.5 m of the Grey Chalk were exposed in the Ash Chalk Pits [564 028], approximately 500 m SW of Filching Manor, Wannock. CJW, RNM, BY, RDL

PLENUS MARLS

Complete sections through the Plenus Marls were noted in the Machine Bottom Pit (p. 68), at Rodmell (Figure 17), and in the Ash Chalk Pits (Figure 17), where they were about 5, 6.3 and 7.2 m thick respectively. Small incomplete sections included those in the banks of tracks near Bridgwick Pit [4327 1142], at Week Lane [4483 1084], and at Firle Bostal [4646 0664], as well as in degraded pits near Firle Beacon [4880 0619] and near New Barn [5117 0441].

An old pit [4983 0525] to the east of the lane 300 m south of Bopeep, near Alciston, formerly provided a good section in the Plenus Marls overlain by the Melbourn Rock (White, 1926; Gaster, 1939). White (1926) noted that this pit yielded good specimens of the belemnite *Actinocamax plenus*. BY, RDL

Upper and Middle Chalk

MELBOURN ROCK

The fully exposed successions in the Lewes and Eastbourne areas are shown graphically in Figure 17. Elsewhere the Melbourn Rock typically forms a good feature and nodular chalk fragments are common in the soil brash. Small incomplete sections include the following:

National Grid reference	Type of exposure	Exposed thickness (m)	Notes
3906 0874	Roadside	1.8	Dip 15°N
4257 0931	Roadside	1.5	Dip 22°N
4327 1142	Trackside	0.6	
4367 0900	Trackside	—	
4481 1082	Roadside	—	
4487 0879	In wooded valley	—	
4496 1023	Roadside	—	
4535 0888	Roadside	—	
4581 0674	Roadside	—	
4583 0689	Old pit	—	
4645 0661	Trackside	0.9	
4907 0595	Trackside	0.6	
4917 0580	Trackside	—	
4983 0525	Old pit	4.6	This thickness includes the nodular Holywell Beds above
5117 0441	Old pit	1.2	
5356 0371	Trackside	—	
5545 0385	Small pit	—	
5620 0316	Trackside	1.8	
5815 0255	Old pit	2.4	

RDL, RNM, BY

BEDS ABOVE THE MELBOURN ROCK

Area west of the River Ouse

The disused chalk-pits on the east side of Offham Hill provide large exposures in the Lewes Nodular Chalk Member (Figure 21). In the pit [4005 1156] immediately west of the Chalk Pit Inn almost the whole member is exposed in a face showing some 60 m of strata. It is difficult to relate the exposed section to the descriptions by Hill (*in* Jukes-Browne and Hill, 1904) and White (1926). Much of the chalk of the *M. cortestudinarium* Zone is inaccessible in this pit but may be seen in pits higher on Offham Hill [3998 1144; 3996 1175]. About 6 m of off-white, thickly bedded chalk with local rubbly nodular beds and regularly spaced beds of flint nodules may be seen in the northern pit, the Cliffe Hardground being present at the base of the section. Specimens of *Micraster normanniae* and early *M. decipiens* are common throughout the sequence. Similar chalk, 17 m thick and at a rather higher position in the *M. cortestudinarium* Zone, is exposed in the more southerly of these pits. The Shoreham Marls are present at the top of this succession; in the lowest 6 m *Cremnoceramus* ex gr. *schloenbachi* is common, associated with sponge beds.

The road cutting for the Lewes by-pass at Juggs Lane [401 093] (Figure 22) formerly exposed 37 m of the Seaford Member dipping north at 20°. Higher beds come in westwards. The lowest beds exposed were notably rich in *Volviceramus* ex gr. *involutus* and *Platyceramus sp.* (i.e. at a level just above the Seven Sisters Flint Band). The Coniacian–Santonian boundary succession with *Cladoceramus*

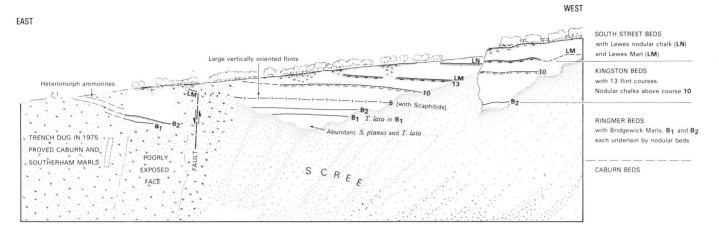

Figure 24   The eastern part of the section at the Bridgwick pit, near Lewes

undulatoplicatus and associated fauna was present 29 m above this level; *C. undulatoplicatus* was very abundant above the third of the distinctive flint bands and the uppermost beds were notably rich in *Conulus albogalerus*. The section in the upper Ranscombe Member near Iford Hill [3992 0640] is shown in Figure 18.

A small pit [4066 0512] near Breaky Bottom exposed part of the Lewes Nodular Chalk Member:

	Thickness m
Moderately well bedded chalk, tending to be rubbly upwards; one course of red nodular flints; local cylindrical flints. *Sternotaxis sp.*, *Micraster* cf. *corbovis*	3.0
Lewes Marl	0.08
Massive nodular chalk with seams of flint nodules	2.4

This is considered to be a condensed sequence, ranging from the Kingston nodular chalks to just above the Lewes Nodular Chalks (Figure 12).

A degraded pit [4109 0477] at Cricketing Bottom showed the following section in the same member (Figure 12):

	Thickness m
Generally well bedded chalk with courses of flint nodules; one tabular flint bed at 1.7 m depth	3.4
Massive chalk with slightly nodular appearance; isolated flint nodules	2.7

In an old pit [4280 0363] near Deans up to 4.3 m of blocky to moderately well bedded chalk in the Seaford Member are exposed with courses of nodular flints at 0.6 to 0.9 m intervals. Up to 7.3 m of similar chalk are exposed in a roadside quarry [4325 0323] at Piddinghoe. Large *Conulus* typical of the highest beds of the Seaford Member are present in the middle of the sequence. Low-angle curved shear planes traverse the face. Both pits lie in the *M. coranguinum* Zone (Gaster, 1929).

Mount Caburn area

About 3.8 m of massive, greyish white chalk with a 0.15 m marly parting were exposed in a small pit [4234 1108] opened in the *Terebratulina lata* Zone 120 m north of the Prince of Wales public house, South Malling. These beds lie in the Ranscombe Griotte Chalk Member. During the course of road widening at the southern end of Malling Street, Lewes, the following section was noted in the same member in the eastern bank of the road [4225 1047]:

	Thickness m
Massive off-white chalk with some marly wisps; locally hard and nodular, especially in the upper part	3.1

	Thickness m
Pale grey marl (one of the New Pit Marls); some evidence of shearing along this bed; sharp base	0.1
Off-white, nodular, rubbly chalk; passing down to	0.15
Massive off-white chalk	1.5

Detailed palaeontological work was not possible and part of the section is now obscured, though the beds seen are assigned to the top of the *T. lata* Zone and contain *Inoceramus* ex gr. *lamarcki*.

The disused pit known as New Pit [4249 1120], 250 m SE of the junction of the A26 and A273 roads north of Lewes, showed up to 34 m of chalk belonging to the Ranscombe Member and the succeeding Lewes Member. In the lower part of the pit regular alternations of thickly bedded chalk and thin irregular marl seams (the New Pit Beds) were exposed (Figure 18). The nodular chalks with flint nodule beds which were recorded in the upper part of the section stood out as a bluff and contained the Southerham Marls and the Caburn Marl (Figure 20); these upper beds belonged to the Lewes Member and included the lower Bridgewick Marl at the top of the exposure. The beds dip at 5°S.

Bridgwick Pit [4310 1117] (Figures 20 and 24), about 550 m east of New Pit, exposed some 18 m of the Lewes Member, dipping at 5°S. The lowest markers exposed were the Bridgewick Marls but beds down to the Southerham Marl were occasionally seen in the western grassy slope. The highest bluff of the generally inaccessible face was formed by the Kingston nodular chalks with digitate annular burrow-form flints. An important '*reussianum* fauna' has been obtained from this locality (p. 59).

Perhaps the largest but unfortunately the most inaccessible sections in the Ranscombe, Lewes and Seaford members in the district may be seen in the disused cement works pits at South Street [4250 1000], Lewes, and at Southerham Works Quarry [4265 0965], Southerham. In the former, known as the Navigation (or Snowdrop) pit, chalk of the *S. planus*, *M. cortestudinarium* and *M. coranguinum* zones is exposed (Figure 21). The Seven Sisters Flint Band forms a prominent marker at the top of the section in the axis of the Caburn Syncline; on the northern limb of the fold the beds dip slightly west of south at up to 6°. The sequence was described by Rhodes (*in* Jukes-Browne and Hill, 1904). It must be emphasised that previous authors (for example White, 1926, p.56) placed the base of the *M. coranguinum* Zone 30 m too low in the succession and other zonal boundaries were also placed erroneously.

The southern end [4265 0940] of the old pit in which Lewes cement works stands is cut through almost the complete thickness of the Ranscombe Griotte Chalk Member dipping at up to 15°NNW. Unfortunately much of the face is now obscured and it was partly so when described by Strahan (1896). White (1926) recorded the most southerly face, providing a strike section of nodular chalk of the *M. labiatus* Zone, followed upwards in the eastern face by more blocky beds high in the same zone; the overlying chalk of the *T. lata* Zone was rather blocky and less nodular. A section visible in 1973, 170 m NE of the level-crossing at the works entrance, and near the top of the zone, showed about 4.5 m of hard, massive, creamy white chalk with a few marl partings and some courses of flint nodules. This pit formerly exposed a local development of phosphatic chalk which was described by Strahan (1896). His section, where the deposit was thickest, appears to have been about 230 m NNE of the level-crossing and reads as follows, with measurements converted to metres:

	Thickness ft in	Thickness m
'Massive chalk with flints		
Flaky white chalk with a few flints and *Holaster planus*; passing down into	4 0	[1.22]
Phosphatic chalk with many small fish-teeth, a few spines of *Cidaris* and some nodules, partly green, partly brown, up to 1½ inch [0.04 m] in diameter		
A sharp line of demarcation	1 6	[0.46]
Hard creamy limestone with calcite in veins and cavities, nodular (some of the nodules being green-coated), lumps of decomposed iron-pyrites	1 6	[0.46]
Hard, white, compact chalk, traversed by branching pipes and thin laminae of phosphatic chalk	3 0	[0.91]
Hard white, compact chalk, with the pipes [burrows] and laminae of phosphatic chalk less abundant and dying away downward'	3 0	[0.91]

Within these beds Strahan also described certain foraminiferal casts composed of a greenish mineral, assumed to be glauconite, and a small flint nodule containing numerous phosphatised foraminifera. Chapman (1896) listed 42 species and varieties of foraminifera and six species of ostracods from this chalk. The deposit was apparently restricted in extent, 18.3 m wide and thinning markedly towards each end of the exposure. The phosphatic rock is composed of numerous shell, bone and tooth fragments, phosphatised foraminiferal casts and tests, fawnish brown coprolites and scattered grains of a dark greenish black mineral, probably glauconite, all set in a pale fawnish grey chalky matrix. The majority of the included fragments are less than 1 mm in size, though occasional tooth, bone and presumed *Inoceramus* shell fragments are up to 5 mm across. Some specimens contain pale fawn coprolites up to 4 cm in diameter.

Strahan equated the phosphatic chalk with the Chalk Rock of the Chilterns and, therefore, took it to mark the base of the Upper Chalk. However, excavations [4267 0943] in 1975 and 1985 revealed a comparable bed which lies considerably below the Bridge-wick Flints (p. 57 and Figures 18 and 20) which are taken to mark this boundary. This exposure showed a fully developed hardground, about 70 cm thick, with a convoluted upper surface strongly mineralised with iron, glauconite and phosphates; the mineralisation was most intense in the top 20 cm. Lustrous chocolate-brown phosphate coatings of burrow walls were conspicuous and bored residual pebbles of mineralised chalk occurred both in the burrow-fills and in overlying sediment. The burrow-fills consisted mainly of re-worked pelletal phosphatic chalk and other debris, and the hardground was overlain by 3 cm of similar sediment. Discrete sponge-rich layers occurred within the hardground and a sparse fauna of poorly-preserved inoceramids was collected from the upper surface. Electron photomicrographs revealed an abundance of calcispheres and foraminifera in the hardground. It is uncertain whether this bed is the same as that described by Strahan or whether it is another lenticular phosphatic chalk at about the

same stratigraphical level, i.e. that of the upper New Pit Marl, but it is clear that Strahan's phosphatic chalk cannot be correlated with the Chalk Rock. The geophysical resistivity logs of the South Street Borehole [4254 0991] and other boreholes support the former correlation. The occurrence of phosphatic chalk at this stratigraphical level is unique in the English Chalk, but may correspond to the relatively weakly phosphatised Tilleul Hardgrounds 1 and 2 of Haute Normandie on the French coast (Kennedy and Juignet, 1974, p.6).

The succeeding Lewes Member chalk is present only in the inaccessible high faces of the Southerham Works Quarry, but the sequence of strata from the Lewes Marl to the Cliffe Hardground was exposed in the cutting south of the road-tunnel portal [4245 1005].

A small and rather degraded pit [4451 1042] 750 m WSW of Glyndebourne Opera House exposed about 4.5 m of well jointed, massive off-white chalk in the Ranscombe Member and of the *T. lata* Zone (White, 1926; Gaster, 1951). Many of the joint faces in this pit showed vertical and horizontal slickensiding.

Chalk also in the Ranscombe Member and belonging to the *M. labiatus* and *T. lata* zones was seen in the Glyndebourne pit [4484 1018] (Figure 18), 650 m WSW of Glyndebourne Farm. The lowest beds were exposed near the entrance to the quarry and these comprised griotte marls and nodular chalks with abundant *Mytiloides* and *Orbirhynchia cuvieri* in association with *Mammites nodosoides*. On the south side of the quarry the beds, which dip south-west, were traced westwards in an almost continuous sequence towards the main western face where the Malling Street Marls were recognised, disturbed by faulting. These marls occur above the lowest of the six Glyndebourne Hardgrounds which form distinctive green and red, iron-stained horizons in this face. These beds are phosphatised and contain weakly phosphatised burrow-fills. Convolute, phosphatised and burrowed top surfaces are characteristic. White (1926, fig.7) depicted the faulting which has affected this sequence and placed the zonal boundary at the level of the second hardground. However, *Terebratulina lata*, *Collignoniceras woolgari* and *Conulus subrotundus* have been collected from the beds below the second hardground and above the Malling Street Marls.

Chalk of the Ranscombe and Lewes members, lying in the upper part of the *T. lata* Zone, was seen in a small pit [4497 0956] about 750 m WNW of Glynde church, where two thin marl seams near the top of a succession of smooth-textured griotte chalks have been named the Glynde Marls (Figure 18). White (1926) described a section in this pit showing two small normal faults downthrowing south and a nearly horizontal thrust fault. The section is now greatly obscured by talus and the faults cannot be seen, though a conspicuous dark marl bed (the lower Glynde Marl), approximately 12 m below the top of the face, showed evidence of horizontal movement; the chalk 0.6 m below and above this bed was considerably shattered. The paired Southerham Marls were seen in the top 1.5 m of the face.

The disused Caburn pit [4473 0891], 275 m east of the summit of Mount Caburn, formerly exposed the whole of the Ranscombe Member at its type locality (Figures 18 and 20). Currently only the uppermost 47 m of the Ranscombe Member and the lowest 16 m of the Lewes Member are well exposed, the lowest 50 m of the section being much obscured. Gaster's zonal map (1951) of the Chalk of this area showed a fault trending roughly north-west, the 'Caburn Fault', running between Mount Caburn and the Caburn pit, but there is no evidence to support the existence of this structure.

Main crop east of the River Ouse

A roadside pit [4340 0483] near Baydean Bottom exposed 4.6 m of massive to moderately well-bedded chalk overlain by 3 m of nodular white chalk in the Ranscombe Member. The chalk here is in the *T. lata* Zone (Gaster, 1929) and appeared to be somewhat disturbed by fracturing and possibly by cambering.

Up to 30 m of white chalk with regular courses of nodular flint (Seaford Member) were exposed in the Tarring Neville quarries [446 034] (Figure 22 and p. 65). The Seven Sisters Flint Band was present at the base of the section and abundant *Volviceramus* ex gr. *involutus* occurred in the basal 5 m. A conspicuous semi-tabular flint horizon (the Tarring Neville Flint) was present 8 m from the bottom of the section. The uppermost 3 m contained conspicuous flint beds—the Michel Dean, Baily's Hill and Flat Hill flints—and *Cladoceramus undulatoplicatus*.

A pit [4437 0584] to the north-east of Itford Hill showed the following section in the Lewes Member, above 3 m of talus:

	Thickness m
Massively bedded white chalk with burrow-fill flints and some thin tabular flint seams	1.8
Marl (the Lewes Marl)	0.1
Hard yellow nodular chalk; *Holaster sp.* at the top	0.5
Massive white chalk with burrow-fill flints; slightly nodular appearance	1.2

This pit lies in the *S. planus* Zone (Gaster, 1929).

An old and much degraded pit [4581 0617] 170 m NNW of the summit of Beddingham Hill exposed 10.7 m of creamy white massive chalk with regularly spaced beds of flint nodules in the Lewes Member. This chalk belongs to the *M. cortestudinarium* Zone according to Gaster's unpublished field notes.

The chalk in the largely degraded Firle pit [4643 0623] 940 m south of Newelm belongs to the Lewes Member and largely to the *S. planus* Zone (Figure 12). The Caburn Marl was seen in the lowest part of the section. The Lewes Marl was exposed in the uppermost part and the adjacent nodular beds contained the distinctive branching tubular Lewes Flints.

Up to 4.6 m of white chalk with flints, of the Lewes Member, were exposed in much degraded pits [4734 0603] immediately south of Firle Plantation. This chalk belongs to the *M. cortestudinarium* Zone (Gaster, 1929). Gaster (in unpublished field notes) noted chalk of the *M. cortestudinarium* Zone in the banks of a dew pond [4789 0562] on the slopes of Well Bottom 730 m WSW of the summit of Firle Beacon. Chalk of the *M. coranguinum* Zone forms the higher slopes of this dry valley and it follows that there is an inlier of *M. cortestudinarium* chalk near the head of Well Bottom. Gaster's zonal mapping of the area (1951) also indicated that many of the dry valleys running south-south-west from the crest of the main Chalk scarp between Beddingham Hill and Firle Beacon might contain inliers of *M. cortestudinarium* Zone chalk towards their heads.

The Bopeep pit [4946 0519], 550 m SW of Bopeep, Alciston, is a conspicuous feature on the scarp face. The highest part of the Ranscombe Member (at least 12 m) and the lowest 17 m of the Lewes Member (Glynde Marl to Lewes Marl) are exposed here, dipping at 6°SW (Figures 18 and 20). The higher beds consist of creamy white, rubbly nodular chalk with occasional flint nodule beds and thin marl beds up to 0.05 m thick. Much of this chalk is rather inaccessible in the pit, though the banks of Bopeep Bostal immediately south of the pit provide exposures of these and higher beds.

Chalk of the Lewes Member, ranging from the Bridgewick Marls to the Lewes Marl, is present in the degraded faces of the Winton pit [5116 0369], about 1100 m south of New Barn (Figure 12). Approximately 4.6 m of white chalk with flint nodule bands were exposed here. The lower part of the pit formerly exposed the junction with the underlying *T. lata* Zone (Gaster, 1939).

East of the Cuckmere River there are few exposures. The large degraded pit [5358 0360] on the south-east side of the road approximately 950 m SW of Wilmington Priory exposed about 7.6 m of thickly bedded off-white chalk of the Ranscombe Member and belonging to the *M. labiatus* Zone. About 3 m of *M. cortestudinarium* Zone blocky white chalk with flints were seen in the old pit [5422 0330] on the summit of Windover Hill, and similar chalk belonging to both this zone and to the underlying zone of *S. planus* (Gaster, 1939) was exposed in the sides of the track [5405 0330] on the western slope of the hill. These exposures lie in the Lewes Member.

RNM,CJW,BY,RDL

CHAPTER 8

Structural geology

MAJOR STRUCTURES

The Lewes district lies close to the southern limit of the Wealden structural area which at the surface shows an anticlinorial structure but at depth has a trough-like form (Lamplugh, 1920). The Mesozoic formational thicknesses are generally greatest in the central Weald and since this axial region was subject both to the greatest subsidence and later to the greatest relative uplift, the term 'inversion structure' has been applied to the region as a whole.

The major deformation of the Wealden Basin probably took place during earth-movements that started in the Cretaceous (Casey, 1961, p.551; Owen, 1971, p.200; Lake and Young, 1978, p.7) and culminated in the Miocene.

The more important minor structures within the anticlinorium have in the past been regarded as subsidiary flexures *en echelon*. It is now considered that the more important subsidiary structures are faults or zones of faulting, which were probably generated at the interface of the Palaeozoic and Mesozoic sediments. The faults observed at the surface are generally confined to the Hastings Beds. In the higher part of the Cretaceous sequence large-amplitude folds are present, and these probably arose because deep-seated fault structures were modified during transmission owing to the lower competence of these strata and the effects of diminished thickness of overburden (Lake, 1975b).

Of the five major structural lineaments recognised in the Weald (Lake, 1975b, fig. 1), two occur in the environs of the Lewes district, namely the fault structures present in the area of the Purbeck inliers and at Pevensey (Figure 25). To the east, in the Hastings district, the former structure shows a well-defined symmetry: it has the form of a faulted anticline or demonstrates horst structures. Folds where present are closely associated with these fault-blocks. In the Lewes district the Dudwell Fault runs approximately axially to the main west–east anticline of the Purbeck inlier, whereas the associated Burwash Common Fault (in the Tunbridge Wells district) has a larger throw and swings north-westwards (Anderson and Bazley, 1971, fig. 2). The Purbeck inlier system is well reflected on the gravity survey map but its possible continuation westwards is conjectural.

The inlier of Tunbridge Wells Sand near Pevensey is dominantly fault-controlled and has typical horst characteristics. Westwards this structure passes into the Arlington Anticline (Figure 25), which shows only minor modifications by faulting. If the deformational intensity was comparable between the two areas, the differences in structural style indicate considerable damping of stresses on transmission through the Wealden beds. Still farther west, the continuation of the Arlington Anticline can be recognised as the Kingston-Beddingham Anticline (Plate 8); the Caburn Syncline, complementary to this structure, lies to the north.

In the ground between the Pevensey and Purbeck inlier structures, which is mostly underlain by Hastings Beds

strata, the structure is dominated by strike-faulting which, with the exception of the Horsebridge Fault (Figure 25), tends to repeat the stratigraphical sequence. In the Weald Clay tract, structures are difficult to detect; those which have been recognised and shown on the 1:50 000 geological map as faults may alternatively be monoclinal disturbances. The strike-faults are regarded as tensional or accommodation structures and subordinate in importance to the lineaments defined above (Lake, 1975b).

The relationship between the surface structure and the deep-seated geology in the southern part of the Weald is a matter of some conjecture. Since the Mesozoic formations thin southwards from the central Weald it is inferred that a basement 'high', the Portsdown–Paris Plage ridge (Taitt and Kent, 1958), is present in the south, but the postulated westward continuation of the shallow basement beneath Paris Plage has yet to be located. Possibly this structure lies close to the coast of Sussex, in which case the Pevensey structure may reflect a major flanking fault. By way of comparison, the gradient of the basement surface on the northern side of the Wealden Basin (about 1 in 13 southward with respect to the base of the Gault) has been ascribed to a tectonic hinge which permitted subsidence of the basin by block-faulting to occur (Shephard-Thorn and others, 1972, p.104).

Within the Lewes and adjoining districts it has already been noted that an appreciable attenuation of certain formations is demonstrable in an overall southward direction: of the combined Purbeck Beds and Ashdown Beds between Fairlight and Pevensey (p.14), of the Ashdown Beds between Dallington and Pevensey (p.16), of the Wadhurst Clay towards Pevensey (p.16), and of the Lower Greensand towards Hampden Park (Eastbourne) (p.36). The first three examples demonstrate a broad regional pattern, whereas the last may be more significant, particularly if considered in terms of the down-dip component (from Ripe on the nose of the Caburn Syncline to Hampden Park, south of the Pevensey fault structure) across the Kingston–Beddingham–Arlington–Pevensey lineament. In contrast, the variations of thickness of various parts of the Chalk sequence (p.66), which are related to a depositional trough at Lewes and flanking swells at Brighton and Eastbourne, appear to have no direct relationship with the strike-orientated structures previously described. In order to account for these sedimentary structural features, Mortimore (1979) has proposed that the swells may be interpreted as the result of continued movement on ancient horsts and grabens in the basement, offset by wrench faults to give a modified orientation to the late Cretaceous structures. However, the absence of significant variations in formational thicknesses across the Kingston–Beddingham Anticline suggests that different structural controls were operative in late Cretaceous times,

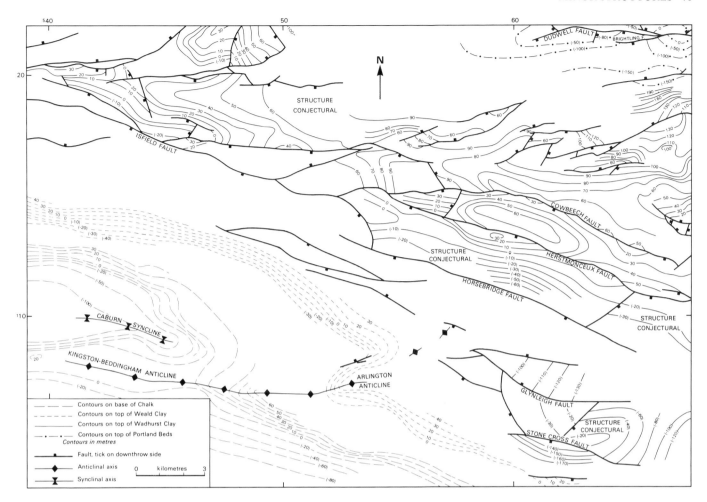

Figure 25 Structural map of the Lewes district

although an overall attenuation in a southward direction is nevertheless apparent in East Sussex. RDL

MINOR STRUCTURES

Borehole cores from the Wealden clay formations have shown that bedding-plane shears are relatively common. Typically the strata between the individual shears show a progressive increase of dip downwards; immediately beneath a shear plane the strata are almost horizontal. This style of deformation has been observed in relatively undisturbed Weald Clay beds (as at the Ripe Borehole [5059 1052]) and in the more consolidated Purbeck Beds in structurally more complex terrain (as at the Broadoak Borehole [6195 2214]). These structures are considered to be an important mechanism whereby deep-seated stresses have been effectively absorbed or partially absorbed and refracted during transmission. Equally importantly they may have served to accommodate stress-fields formed between major structures.

The presence of bedding-plane shears within argillaceous strata may have provided an agency whereby valley-bulging took place (p. 89). Under periglacial conditions it is probable that within the permafrost zone the shear planes were reactivated and facilitated mass-movement into the valleys.

The fractures in the Chalk fall into three types: minor faults, shear planes and joints (Mortimore, 1979). The shear planes are distinguished from the minor faults by the absence of any noticeable offset of the beds.

Minor faults are quite common and many are associated with fracture zones, gouge, slickensiding or solution cavities. Calcreted surfaces have also been recognised locally. The throws noted range between 0.6 and 7 m but are typically about 1 m. Two well-developed N–S-trending vertical faults with throws of about 1 m are visible in Tarring Neville quarry [4456 0350]; solution cavities along their surfaces are now infilled with chalk and flint fragments and some sands, silts and brown clays. Analysis of the minor faults in the Mount Caburn quarries shows a tendency for two sets of inclined faults to be dominant, one trending at around 330° and the other at 060°; the inclinations are typically steep, between 70° and 80°.

Shear planes are ubiquitous and can be divided into vertical sets trending at 050° and 330° and four inclined sets. The shear planes have slickensided surfaces and tend to undulate continuously. They occur most commonly towards the base of the Ranscombe Member.

The joints show a tendency to form a regular pattern in terms of orientation (Mortimore, 1979) and are most common in the weathered parts of the chalk. However, major open fissures may occur in the chalk near the river gaps as a result of stress relief towards the valleys. Examples were noted in the Cuilfail Tunnel [4230 1020] and its approaches.

RDL,RNM

EARTHQUAKES

In his 'History of British Earthquakes' Davison (1924) recorded three minor earthquakes in the Lewes area. On October 25th 1734 a 'strong' earthquake affected Portsmouth, Chichester and Lewes at around 3.50 a.m.; two shocks were felt at Lewes. A second earthquake occurred on April 30th 1864 at 11.06 p.m. The area affected was 16 km in diameter, with the centre situated some 6.4 km north of Lewes. A noise '... like that of hail driven against a window ... heard after midnight at Lewes and Maresfield' was probably related to an associated shock. The third earthquake, at 1.27 a.m. on August 21st 1864, was felt at Lewes and for up to 24 km around the town. This tremor was described as 'a shock, accompanied by a rumbling sound like the passing of a heavily-laden waggon over a rough stone surface.'

All are estimated to have been of an intensity of 5 on the Richter scale. They were presumably related to movements along structures in the sub-Mesozoic basement, though the nature and position of these features are unknown.   BY

DETAILS

Major structures

Information on individual faults can be obtained from the 1:50 000 geological map and from the structural diagram (Figure 25). The following details therefore include only data additional to the above and notes on the more important structures. In formations where there is limited stratigraphical control, particularly in the Ashdown Beds, it is probable that many structures remain undetected.

The Brightling Borehole [6725 2182] proved a reversed fault in Liassic strata at 1137 m depth (p. 7). The fault throws about 180 m and is probably associated with the Burwash Common Fault but may lose some of its effect in the Kimmeridge Clay as suggested by Howitt (1964, p.101).

The Dudwell Fault runs axially to the anticline which causes the Purbeck Beds to crop out in the Dudwell valley. True dips for this fold of the order of 8° to the north and 2° to the south are indicated by sub-surface data. The observed dip values at the surface in this area have probably been affected by local superficial movements. The Broadoak Borehole [6195 2214] showed a northward dip of about 5° for undisturbed strata (Lake and Holliday, 1978).

The Cowbeech and Herstmonceux faults flank an anticlinal structure whose axis runs through Jarvis's Wood. The fold has an abrupt westward termination. It probably owes its origin to tilting between the faults and terminal-bending against the Herstmonceux Fault. The Grove Hill Borehole [6008 1359] was sunk on this structure in 1937 in search of hydrocarbons (p.5).

The Horsebridge Fault forms the northern boundary of the Weald Clay in the eastern part of the district. Its precise position and nature are somewhat conjectural, since over much of its length it throws silty clays and clayey silts in the upper part of the Tun-

bridge Wells Sand against Weald Clay. Excavations [5854 1146] near Horsebridge Mill, however, showed fine-grained sands of the former formation in close proximity to buff Weald Clay. In the Burgh Hill area [542 127] the Tunbridge Wells Sand consists mainly of silt and the lithological contrast with the Weald Clay is greater. The throw of the fault is estimated to be about 115 m down to the south in the Horsebridge area.

In the eastern part of the Weald Clay tract, the tracing of marker clay beds has enabled some details of the structure to be elucidated. In the deeply wooded area to the west of Hailsham, however, the continuity of the beds and the inferred structure is uncertain. Minor SW–NE-trending anticlines have been recognised in the vicinity of Michelham Priory. Strong inflections of the outcrop of the clay marker bed in the woodland south-west of Hailsham are assumed to have been caused by faulting; they may equally have resulted from monoclinal folding. The Southdown or Keymer Tileries [583 069] formerly showed evidence of flexuring of the Bed 4 marker clay. In the area [572 054] to the north-east of Wootton Manor, the form of the outcrop of the Lower Greensand suggests the presence of a south-facing monocline, which may be related to the faulting to the east.

The Glynleigh and Stone Cross faults are the major bounding faults of the Pevensey inlier. The structure shown in Figure 25 is conjectural because part of the outcrop area is obscured beneath alluvium and because stratigraphical control is poor in the vicinity of the Stone Cross Fault. The Pevensey No. 1 Borehole [about 626 054] is now considered to have penetrated faulted ground; the stratigraphy of the fault block to the north-west of this site was resolved by the Glynleigh Borehole [6085 0637] (p. 103). Bituminous seepages are apparently closely associated with the faulting in this area (p. 32).

Minor structures

In the Ripe Borehole [5059 1052], clearly defined bedding-plane shears and zones of listric surfaces were noted below 20.65 m to the bottom of the borehole at 214 m. The listric surfaces typically affected the more homogeneous mudstones. The dip varied between 0° and 15° in an oscillatory fashion down the hole, showing a tendency to be greater below 200 m. It was not possible to record the dip variations in detail because of the extent of shearing in many of the beds. It was noted, however, that the dip increased downwards from 8° to 15° above a zone of listric surfaces at 61 m. Below this level a value of 7° was recorded. Similarly dip values above and below a zone of listric surfaces at 65 m were 7° and 3° respectively. Dip values within beds affected by listric surfaces between 194.50 and 195.50 m varied from less than 5° to 15°. Only a slight dip was noted immediately below the latter depth.

In the Broadoak Borehole, shear zones occurred at 17.50, 25.64 m (at 45°), 26.34 m (where the dip increased to 45°), 30.0 m (associated with variable dips), 34.09 and 71.85 m. Listric surfaces affected the softer lithologies throughout the sequence. The dip varied from 10° to 15° between the surface and 25.64 m. Below this, values from 8° to 10° were observed, increasing to 18° adjacent to the shear zone at 34.09 m. Dips of less than 5° were noted below this level, increasing to 7° around a depth of 80 m and then reducing below to 3°. Dips recorded above and below the Gypsiferous Beds (which were present from 118 to 134 m) were 5° and 7° respectively. There was evidence of open fracturing down to 28 m; it is probable therefore that the steeper dips in the higher part of the sequence are due to superficial effects rather than tectonic movements.   RDL

CHAPTER 9

Quaternary

INTRODUCTION

At times during the Pleistocene glaciations permanent fields of snow and ice capped the higher ground of the Lewes district and the ground was frozen to a considerable depth over the whole area. It is possible that sea-ice extended from the North Sea basin into the area of the English Channel and consequently modified the drainage system of the southern Weald. Under these conditions erosional and depositional features were produced, only relicts of which are evident at the present day; the crags of the High Weald, the coombes and dry valleys of the Chalk downland, and extensive solifluction deposits were probably formed during these times. Cambering, land-slipping and valley-bulging of strata in valley locations also took place mainly in cold periods.

Variations in the base-level of the drainage systems during and since Pleistocene times caused alternate downcutting and aggradation, giving rise to a series of river terraces. Cryoturbation structures in the older river deposits reflect subsequent periods of permafrost conditions. The base of the alluvial deposits of the major rivers is graded to a lower sea level than that now prevailing and relatively recent submergence in Flandrian times has caused the infilling of the valley bottoms with gravels and alluvial mud. Flooding of the clay vale north of Pevensey in the same time period caused the development of extensive estuarine flats in the area now known as Pevensey Levels. Subsequent damming by the migration of storm-beach deposits along the present coastal strip and siltation gave rise to the low-lying marshland in this area. At the present day, rivers are building up alluvial deposits on their flood-plains and deposits of peat and silt are accumulating at spring-lines.

Precise palaentological evidence for dating Pleistocene and Recent events is rare in this district so that only a broad empirical description will be attempted. The denudation history can to a certain extent be inferred from the Wealden area in general. By analogy with the Brighton district (Young and Lake, in preparation), it is possible that the Second Terrace of the River Ouse is equivalent to the Second Terrace of the River Adur, which is thought to be broadly contemporaneous with the Brighton Raised Beach and of Ipswichian age. The various Head deposits are probably largely of Devensian age, whereas the higher terraces, the Clay-with-flints and Older Drift Deposits date back to at least the Wolstonian and beyond. RDL

CLAY-WITH-FLINTS AND RELICT TERTIARY DEPOSITS

Large fragments of conglomerate, composed of green-coated subangular flints and quartz pebbles in a dark reddish brown sandstone matrix, and of ferruginous sandstone largely derived from the Woolwich and Reading Beds of Tertiary (Palaeocene) age are common in the superficial deposits on the South Downs. Sands from the same beds infill pipes and other cavities in the Chalk. Hard cemented sandstones known as sarsens or graywethers, which are probably relics of a Tertiary or early Pleistocene silcreted layer, are also present as isolated blocks on the higher Chalk downland. Indeed, minor relict patches of Tertiary sediments may be concealed by the Clay-with-flints, which characteristically consists of reddish brown, stiff, sandy clay with flint fragments and Tertiary debris such as that described above. The flint content in the Clay-with-flints is variable. Individual flints are unworn to well-rounded and show a distinctive 'bleached' patina locally spotted with black manganese oxide. Local sandy intercalations are present within the overall clay matrix. The deposits formerly extended as an almost continuous sheet, but now only dissected remnants remain, about 1 to 2 m thick, generally on the high part of the Chalk dip-slope.

Flinty clays occurring at lower topographic levels are regarded as solifluated derivatives of the Clay-with-flints *sensu stricto*. At the margins of the Clay-with-flints outcrops leaching of the finer matrix material gives rise to fringing spreads of loose flint gravels.

The Clay-with-flints is regarded as largely derived from former Tertiary deposits, with only a minor contribution afforded directly by the dissolution of the Chalk. Pepper (1973) has reviewed the literature concerning these deposits. RDL

DETAILS

White (1926, p.62) noted an exposure of flinty loam, almost certainly Clay-with-flints, in the top of a solution pipe exposed in a track south-east of the Grand Stand at the now disused Lewes Race Course. He suggested that the flint pebbles might have been derived from the basement bed of the London Clay. The exposure cannot now be located.

A very small patch of sandy Clay-with-flints was formerly visible in the old diggings on the summit of Cliffe Hill [4340 1072]. According to White (1926, p.62) the flinty loam here contained large pockets of red and orange, generally fine-grained sand, with some rounded to subangular quartz grains up to 2 mm in diameter. The sand contained fragments of ferruginous sandstone and clay-ironstone. White compared this sandy deposit with the Woolwich and Reading Beds occurring in a small outlier at Falmer, west of the Lewes district (White, 1924). Gaster (1937b) described this material as being piped into the chalk of the *M. coranguinum* Zone and recorded having found a green-coated flint. Little can be seen today and the deposit is too small to appear on the map.

Relatively small patches of Clay-with-flints cap the higher hills along the main South Downs escarpment between the River Ouse and the Cuckmere River and there are scattered patches on parts of the dip-slope ridges to the south. The deposit appears to be generally less than 1.5 m in thickness and though there are few exposures the presence of Clay-with-flints is easily detected in arable land by the reddish brown clay soil with many flints. Patches of gorse locally mark the outcrop of the deposit in unploughed pasture land, for example on Firle Beacon [4850 0580].

Approximately 1.2 m of typical Clay-with-flints was noted overlying massive white chalk of the *M. coranguinum* Zone in a small pit [4615 0581] 350 m SE of the summit of Beddingham Hill. Rather larger exposures of typical Clay-with-flints were seen in old pits [4932 0507; 4935 0511] a few metres west of the car park on Bostal Hill above the Bopeep chalk-pit. In both of these sections the material consisted of stiff reddish brown clay with some sand grains, abundant bleached flints and some fragments of coarse-grained brown ferruginous sandstone containing flint pebbles. In the first section the Clay-with-flints appeared to be piped into a rather broken, possibly frost-shattered chalk surface, and traces of cryoturbation were detected in the Clay-with-flints.

Silty and sandy Clay-with-flints was noted in the south side of the track [5478 0339] on Wilmington Hill, 500 m east of the Long Man. Brown sandy loam was seen in the 1.8-m sides of an old pit [5489 0331] about 120 m south of the triangulation pillar on Wilmington Hill. Blocks of dark brown, coarse, ferruginous sandstone are abundant in the soil covering this Clay-with-flints outcrop. Coarse brown sandstone fragments of this type are common in most patches of Clay-with-flints and in the soil developed on their outcrops.   BY

OLDER DRIFT DEPOSITS

The Older Drift Deposits are not differentiated on the published map. Extensive spreads of flinty drift are present in the area reaching from the foot of the Downs on to the outcrop of the Tunbridge Wells Sand around Uckfield and Chiddingly. These are probably largely of solifluction origin. Flints also occur within the highest river terrace deposits and were presumably derived from much older deposits pre-dating the (30-m) Fourth Terrace of the River Ouse. They are also present as isolated clasts within the top soil over the outcrop areas of the Weald Clay and part of the Tunbridge Wells Sand. Characteristically the flints are buff to dark brown, with a waxy patina and they display small circular surface pitting ('pot-lid' structure) and slightly abraded fracture surfaces. The examples observed in the river gravels show a greater degree of rounding (Kirkaldy and Bull, 1940, p.129).

Quartzite pebbles of exotic nature have been recorded in association with the flint deposits within terrace gravels at Sharpsbridge, Isfield and Hellingly (Kirkaldy and Bull, 1940, pp.131–134), and within pipe-infill deposits in the chalk of the Ouse Gap (Dr G. A. Kellaway, personal communication). From the limited evidence available it would appear that some form of glacial origin for some of these deposits cannot be excluded.

Kirkaldy and Bull (1940, p.133) considered the gravels at Barkham [4392 2161] (the 'Piltdown Man' locality) to be a unique deposit. In the present context, these gravels are equated with the Third Terrace of the River Ouse (see p. 82).   RDL

DETAILS

Abundant large angular flints were noted in the soil near High House Farm [415 187]. The soils associated with small patches of Fourth Terrace at Flinty Bank [426 180] in Oldpark Wood and eastwards were noted to be appreciably flinty. A similar association

was observed to the east of Broomlye Farm [431 198]. Angular brown flints were noted in the soil in the area [405 162] to the north-west of Gallybird Hall. Kirkaldy and Bull (1940) recorded other localities in the Ouse valley where flints generally occur within the deposits of the lowest three terraces (see also p. 82).   CRB

Brown angular flints occur in association with Head deposits to the east of Shelley's Folly [404 150] and in the area [415 146] to the south of Curd's Farm.

Brown flints with a characteristic thick patina and marked surface pitting ('pot-lid' structure) are common in the soil over much of the Weald Clay outcrop between the River Ouse and Laughton. Such flints are locally very abundant in the loamy soil covering the Tunbridge Wells Sand outcrop in the fields surrounding Halland Park Farm, East Hoathly [5110 1590], but there is no mappable thickness of drift material on the high ground here (up to 55 m above OD). Similar flints also occur in the soil covering the Head deposits mapped on the lower ground west and south-west of the farm, though it is not known whether they actually occur within this Head. The concentration of flints on the higher ground suggests that they represent a remanié of an older drift deposit, now almost totally removed by erosion; the flints found at lower elevations presumably reached their present position by soil creep.

Brown 'pot-lid' flints are common in the soil in the area [520 148] to the east of Park Corner and near Muddles Green [545 133]. Similar flints were noted in the soil near Deanland Wood [527 120] and around Tile Barn [542 107].

Brown, thickly patinated flints are very abundant in the soil covering the Weald Clay and the Lower Greensand outlier above 30 m OD west of High Barn [5462 0948], Upper Dicker, though no significant thickness of Head was mapped here. Flints are extremely common in the soil capping the ridge top at above 30 m OD immediately east of Stonery Farm, Selmeston [5215 0695].

Flints are locally common in the area of the Weald Clay outcrop between Hailsham and Polegate and in the soil near Hankham. They also occur in the terrace deposits of the Cuckmere River (p. 83) and in the basal part of the alluvium of the Pevensey Levels (p. 85).   BY, RDL, CRB

HEAD

In this memoir the various drift deposits which cannot be definitely ascribed to an aqueous mode of deposition are termed Head. They include the Dry Valley and Nailbourne Deposits. The classification used is based primarily on lithological characteristics. Solifluction, especially in cold (glacial) episodes, has been the prime mode of transport so that the lithologies reflect the local parent material, although wind-blown deposits may have been incorporated to a minor degree.

On tracing the Head deposits northwards from the Chalk escarpment, a grading of lithologies is observed in places, from flinty marl through flinty loams and clays with subordinate chalk content, to relatively flint-free loams. In other cases a distinct demarcation of lithological types is evident, which may indicate chronologically separate solifluction phases. Because of the variable disposition of Head deposits an arbitrary minimum thickness of 1 to 1.5 m has been adopted for mapping purposes.

The bed-rock underlying the solifluction deposits has commonly been destructured by permafrost conditions. In places the Chalk exhibits degrees of frost-shattering so intense that a soft rubbly marl has been formed. The Gault similarly presents a structureless aspect and secondary 'race' is commonly present. Both the bed-rock and the Head deposits may show cryoturbation structures reflecting a later phase of permafrost conditions.

The dry valley deposits of the Chalk downland are distinguished purely on topographic criteria. Previously White (1926) has described these Head deposits as Valley Gravel, Coombe Rock and Talus. Gradations between river terrace deposits and their re-mobilised derivatives are observed in the district and for convenience these are detailed in the section on River Gravels (p.81).

Included with this group are the Coombe Deposits, that is a type of drift found in association with the coombes of the Chalk escarpment. These solifluction deposits are white to pale grey and are composed of angular chalk rubble with subordinate angular flints in a matrix of chalk mud. Characteristically they are poorly sorted, roughly stratified and may show an upper zone of decalcification. Locally re-precipitation of calcium carbonate has later cemented the deposits into a hard massive rock termed 'Coombe Rock'.

The deposits which are loosely termed 'Dry Valley and Nailbourne Deposits' represent successive periods of solifluction in the dry valleys of the dip-slope. These deposits include solifluction material identical to the Coombe Deposits

sensu stricto, in addition to variable amounts of water-laid calcareous flinty loams or flint gravels formed when permafrost conditions facilitated surface run-off.

In terms of geomorphology, the typical coombes and dry valleys are well developed in the Lewes district (Plate 7). The coombes show the familiar profiles of steep-sided re-entrants in the escarpment, and the drift deposits in them spread out laterally at the foot of the scarp and have been subsequently removed by erosion to varying degrees. The dry valleys of the Chalk dip-slope, which presumably reflect an older drainage system sustained by the formerly more extensive cover of Tertiary deposits, for the most part demonstrate the anticipated consequent drainage pattern. In the area to the south-west of The Brooks [420 080], however, a complex drainage system is presented, showing vestiges of a trellis pattern with an overall preference for an almost obsequent north-easterly trend for the major valleys. One dry valley, Whiteway Bottom, rises within a short distance of the Chalk scarp, below Iford Hill, and after following a SE-trending reticulate course to below Fore Hill [412 048] trends northeast to join the Ouse valley at Rodmell. Just above the confluences of the tributary valleys in this particular system, certain tributary floors show an increase of grade, suggesting

Plate 7 Coombes in the Chalk escarpment of the South Downs near Folkington
(A 12284)

either immaturity of drainage development or more probably competition between aqueous deposition/erosion and solifluction effects. In the upper stretches of the dry valleys other irregularities of profile are also noted, possibly accentuated by the effects of sink-hole formation.   RDL

DETAILS

Head on the Hastings Beds

Narrow belts of Head have been mapped along the lower slopes of many of the valleys crossing the Hastings Beds outcrop. Much of this material, which consists generally of brown to dark brown silty and sandy loams, is undoubtedly a product of hill wash and has been preferentially generated at spring-lines. In a number of valleys the Head deposits are more widely distributed on one side of the valley — usually the east-facing and more gentle slope — than the other. Subsequent erosion has probably removed the Head from the steeper valley side. A minor amount of the Head may have an aeolian origin.   BY

Head on the Weald Clay and Lower Greensand

West of the River Ouse, deposits of Head on the northern part of the Weald Clay outcrop consist generally of silty loams associated with present-day streams and most probably have resulted from the redistribution of older drift deposits. South of the tributary of the Ouse which flows north of Barcombe Cross [421 158], more clay and gravel are present in the extensive Head deposits, which are generally of a loamy consistency. Head deposits adjacent to the terrace deposits of the Ouse valley consist of orange-brown sandy loams and are for the most part derived from the formerly more extensive higher terraces.

Topographic rises [402 151; 408 150] north of Cooksbridge (see also p. 36) have surface cappings of glauconitic clayey loam and yellow-brown clayey silt, and locally flints are abundant in the soil. These deposits were largely derived by solifluction from the Lower Greensand. Extensive spreads of Head in the valley which runs near Cooksbridge [402 139] display a variety of lithologies, ranging from ochreous sandy and silty loams to ochreous-mottled pale grey sandy clay and mottled clayey loam. Much of this deposit has apparently been derived by *in situ* weathering and localised solifluction from the underlying formation. Flinty deposits seem to be mainly restricted to the south slope of the valley. North of Lower Tulleyswells Farm a ditch [3990 1403] exposed 1 m of silty loam with a basal seam of angular flints overlying Lower Greensand.

East of Norlington [447 132] Head is present over much of the Lower Greensand outcrop, the typical lithology being ochreous silty loam with locally a minor glauconite content derived from the underlying formation. In some places a strong ferruginous impregnation imparts a gritty texture to the deposit. Between Ringmer [450 125] and Wilmington [545 045], and particularly in the Ripe area, a surface wash of brown flints is commonly present; this Head may be of considerable antiquity, since there is no apparent source from which the flints could have been derived. The deposits around Ripe are in places banked against 'islands' of Lower Greensand. A temporary section at Curls Farm [5075 1016] in Ripe showed 1.5 m of loam with carbonised roots locally penetrating the full sequence. A borehole [506 105] at Ripe proved 3.5 m of firm brown and grey mottled silty sandy clay (Head) with carbonaceous matter overlying 2.6 m of dark silty sand with clay (Lower Greensand).   RDL

In places in the Selmeston [510 070] and Arlington [544 074] areas Head covers high ground in a situation where its mode of origin is obscure. A small patch of rather flinty Head caps the small hill above 15 m OD south of Polhill's Farm [5309 0688], Arlington.

White (1926) gave the following description of sections of Head, no longer visible, in the now-abandoned Cuckmere Brickworks pit [525 071] near Berwick Station: 'Excavations in the south-western part of the yard give a nearly complete cross-section of a shallow channel, about 50 yds. [45.72 m] wide and 12 ft. [3.66 m] deep, filled with roughly bedded and laminated sandy loam, containing much rusty glauconite, and resembling the partly weathered portions of the adjacent Lower Greensand, from which it is in a great measure derived. Brown and white little-worn flints are irregularly distributed through the loam, and are often disposed in short seams and clusters'.

'According to Mr. W. S. Laker, the manager of the brickworks, the channel formerly continued north-eastward across the main brick-pit, broadening out to about 100 yds. [90 m] by the kilns, and decreasing in depth to 3 or 4 ft. [about 1 m] in that direction. In the rising ground south-west of the existing section the channel must shallow more quickly, for Mr. Laker states that it was not to be seen in the neighbouring railway-cutting which is 'all in Sussex Greensand'. The channel follows the pronounced slope of the ground in its vicinity, and is not in the line of any existing watercourse. The material filling it is a local wash, accumulated under abnormal weather conditions, and probably contemporary with the Coombe Rock'.   BY

In the Polegate area Topley (1875, p.204) described a 'stiff clayey wash, almost like shale in places, but containing chalk flints, 8 feet [2.4 m] thick in one place.' Topley described the location as west and south-west of the railway station, which at that time (1868) was west of the former junction with the Hailsham branch-line. Similar material seen in 1967 in a temporary excavation [5787 0446] near the recreation ground was 1.8 m thick and contained well-rounded flints about 1 cm in diameter.

The peripheral slopes of the Pevensey Levels are notably free from drift deposits. Small areas of silty loam occur around Downash [601 078] and the 'island' on which The Lewens [6096 0738] stands is covered by similar material.   RDL

Head on the Gault and Chalk

The Gault outcrop between Glynde [4570 0890] and Laughton [5020 1310] is locally covered by a mantle of Head varying between brown silty loam and brown silty clay, generally similar to much of the Head covering parts of the Weald Clay and Lower Greensand outcrops. Isolated patches of this material on relatively high ground south of Broyle Place [4780 1285] and north-west of Hall Court Farm [4978 0938], Ripe, suggest a possible origin as river terrace.

A newly cleared ditch [4930 0860] on the east side of the road 500 m east of Bushy Lodge, near West Firle, exposed up to 0.75 m of fawnish grey clayey loam with a discontinuous layer of angular to subangular flints at the base, resting on pale grey Gault clay. The flints had a battered appearance and locally the base of the flint layer was seen to be contorted by cryoturbation. Owing to its small thickness this clayey Head was not mapped.   BY

In the Ripe area auger evidence suggests that there is a distinct demarcation between the homogeneous loams on the Lower Greensand crop and the flinty Head overlying the Gault.

A borehole [4187 1034] at the site of the recently built bridge-crossing in Lewes found 7.32 m of Coombe Deposits beneath 5.03 m of alluvium and 1.37 m of made ground (see p. 86). At the north-western edge of The Brooks [around 4092 0938] boreholes proved up to 15 m of Coombe Deposits beneath thin alluvium and, farther east, a site investigation borehole [4218 0917] in The Brooks encountered 1.5 m of Coombe Deposits, resting on Lower Chalk and overlain by 14.4 m of alluvium and made ground (p. 86).

White (1926, p.69) described a road-cutting section [4145 0640] near Northease Farm where 'gravel and gravelly brick-earth [of the Third Terrace (p. 82)] are capped by a few feet of Coombe Rock, consisting of coarse chalk and flint rubble, passing up into marl with

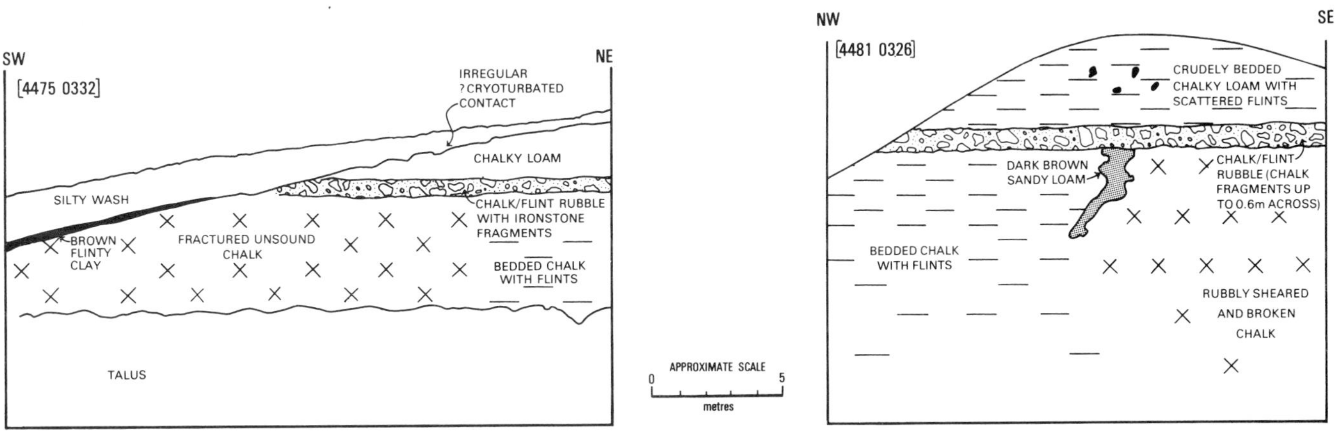

Figure 26   Sketches of sections in Drift deposits at Tarring Neville chalk-pits

small pellets of chalk. The base of the Coombe Rock is uneven and sharply defined'.

The Rodmell Cement Works pits formerly exposed Coombe Deposits; a section [4400 0595] in 1973 showed up to 5 m of crudely bedded pale brown chalky loam resting with an irregular undulating base on extensively frost-shattered Lower Chalk.

Sections at the Tarring Neville chalk-pits [448 033] show the relationship between bedrock, dry-valley deposits and more recent talus particularly well (Figure 26; Ellis, 1985). Minor fault-planes in the chalk are locally veneered with sands and clays.   RDL

Tongues of chalky rubble occupying the coombes in the South Downs escarpment between Beddingham Hill [4587 0600] and Firle Beacon [4857 0592] unite at the foot of the scarp to give an extensive blanket of chalky Head, often described as Coombe Deposits. In places this material appears to have been decalcified and is seen as a brown, silty or clayey, flinty loam, especially in the area around Preston Court [4590 0760]. This Head extends for some distance from the foot of the scarp and caps the Gault outcrop north of Wick Street [4740 0800]. The greater part of the Head covering the Gault here appears to be totally decalcified and seems to grade into, or merge with, the brown, silty, loamy Head of much of the Gault outcrop. The presence of this Head as a capping to the low Gault clay rises in this area and its absence from the lower slopes suggest that erosion of part of the deposit took place during a late Pleistocene phase of downcutting.

Narrow tongues of chalky Head occur in all of the coombes in the escarpment between Firle Beacon and Wilmington Hill [5484 0343] and unite to give more extensive spreads of chalky Head on the lower ground in the Alciston [5055 0555] and Berwick [5190 0510] areas. A passage from the typical chalky 'Coombe Deposits' lithology to decalcified brown silty clayey Head can be noted in places as the deposits are followed away from the scarp.

Narrow belts of thin dry-valley 'alluvium' have been mapped in all the dry valleys of the Chalk dip-slope between the River Ouse and the Cuckmere River. These deposits typically consist of dark brown silty clay, with abundant large unworn flints. Patches of soliflucted chalk rubble, identical to the 'Coombe Deposits' of the scarp face, occur but it has not been possible to map these. Scattered fragments of chalk rubble cemented into hard 'Coombe Rock' by secondary deposits of columnar, cream, crystalline calcite were noted on the higher slopes [4525 0545] on the western side of Well Bottom, south-west of Beddingham Hill.   BY

RIVER GRAVELS

The River Ouse valley contains a series of four terraces of river gravel, as well as alluvium. Distinctive terrace deposits appear to be lacking in the Ash Bourne drainage system north and east of Herstmonceux, and only two levels of terrace have been recognised in the Cuckmere River system. The gravels consist almost entirely of local Wealden ironstone and sandstone debris, which is progressively diluted by flint material in the terraces occurring within the gaps in the Chalk. As noted previously (p. 78), flint and a little quartzite material from older drift deposits have been incorporated in the gravels on the Lower Cretaceous outcrop. In many places the terrace deposits appear to have been partly remobilised by the effects of solifluction. Distinction between *in situ* terrace deposits, reworked terrace deposits and Head deposits is in some instances impossible, owing to the lack of exposure and diagnostic lithological characters; this difficulty particularly applies to deposits of loamy consistency which may present variable topographic features.

The vertical distribution of the terrace deposits in the Ouse valley relative to present-day alluvium level is shown in Figure 27. Owing to poor demarcation of individual terrace units, presumably caused by remobilisation, only those locations where a 'back feature' to the terrace has been recognised have been used for correlative purposes in the figure. This diagram does not indicate the extent of the various deposits.

The higher terraces (Fourth and Third) possess back-feature levels ranging about 30 and 21 m above the alluvium respectively, and consist dominantly of rudaceous material. Although these deposits locally merge, they remain more or less separate from the lower terraces (Second and First) which generally have levels ranging from 10 to 15 m and about 5 m respectively above the alluvium and mostly consist, superficially at least, of loams. These latter terraces are generally poorly defined, commonly merging together and being distinguished with difficulty from certain of the Head deposits.

It is generally believed that the deposition of river terraces was related to changes in the base level of the drainage system, which caused alternate aggradation and downcutting. However, Worssam (1973) has reviewed the denudational history of the Weald and proposed that some river terraces were subject to the controlling influence of climate rather than of sea level. This theory would account for

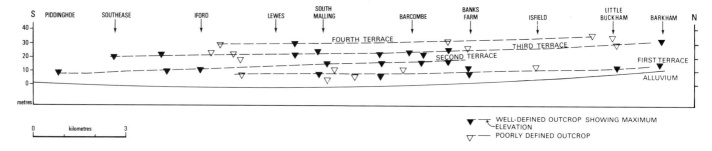

Figure 27 Profile showing the terrace deposits of the River Ouse

broadly contemporaneous terraces in separate drainage basins having different ranges of altitude. On the other hand, the levels of lower terraces which occur near the seaward end of a valley would be determined by oscillations of Pleistocene sea level and this latter thesis is considered applicable over much of the Lewes district.

Many of the river gravels in the Lewes district have been dug on a small scale in the past but none is of economic importance today.   RDL

DETAILS

River Ouse valley

The highest (Fourth) terrace occurs as isolated patches north of Isfield, and Barcombe Cross is partly situated on a more extensive spread. Other outcrops are present north of Lewes Castle [4116 1046] and east of Swanborough Manor [402 077]. White (1926, p.69) described a section at Southease crossroads [422 053] as showing '20 ft. [6.1 m] of red loamy gravel (mainly angular flints) occupying an exceptionally large pipe, of which neither the bottom nor the southern side is seen.' This location is at 30 m above OD and the exposure could represent the relict river-cliff of the Fourth Terrace. White (1926, p.70) also described 'piped remnants of the same terrace-gravel ... in the bluff at Piddinghoe', although this site [about 435 030] is topographically much lower and the deposit is now assigned to the Second Terrace.

The Third Terrace has a wider distribution than the Fourth; locally it appears to merge with its higher predecessor. The well-known Piltdown gravel is part of the Third Terrace; this deposit, which was formerly exposed in the now-degraded pit [4392 2161] at Barkham Manor, was described by Dawson and Woodward (1914, p.82) as:

		Thickness m
1	Surface soil with occasional iron-stained subangular flints, flint implements of all ages and pottery	0.30
2	Pale-yellow sandy loam with small lenticular patches of dark ironstone-gravel and iron-stained subangular flints. One Palaeolithic-worked flint was found in the middle of this bed	0.76
3	Dark brown ferruginous gravel, with subangular flints and tabular ironstone, Pliocene rolled fossils, and *Eoanthropus* remains, *Castor* etc., 'Eoliths' and one worked flint. Floor covered with depressions	0.46
4	Pale-yellow finely-divided clay and sand, forming a mud reconstructed from the underlying strata. Certain subangular flints occur, bigger than those in the overlying beds	0.20

Undisturbed strata of the Tunbridge Wells Sand ...

The discovery of *Eoanthropus dawsoni*, which at that time (1912–17) was regarded as a highly significant 'missing-link' in the study of human evolution, has been subsequently discredited (Oakley, 1955) as a complete hoax. None of the humanoid remains came from Piltdown and more than half the fossils have been shown to have been artificially stained to match the gravels. It is debatable, therefore, whether any reliance can be placed on either the faunal list or the artefacts presented by Dawson and Woodward (1913; 1914) and further discussed by White (1926, pp.64–67). Bed 2 showed an uneven, disturbed base and may be regarded as Head, derived in part from the bed below. 'The junction of Beds 3 and 4 is not always ... sharply defined' (White, 1926, p.65).

South of Barkham, the next deposits referable to the Third Terrace are the gravels with flints south-west of Sharpsbridge [437 205] and gravels at Banks Farm [431 162], Barcombe, and at North End [412 132]. Godwin-Austen (1851, p.288) recorded remains of 'large mammalia' and elephantine remains from gravels at Barcombe.

Elsden (1887, p.646) described a section in a railway-cutting near Lewes [believed to be 4130 1045] as showing very angular flints with some Tertiary pebbles but scarcely any signs of stratification. This deposit and one at a similar level on the other side of the valley, at South Malling, are taken as representing the Third Terrace.

South of Lewes, degraded patches of Third Terrace gravel occur on the Rises, near Swanborough Nurseries [403 083] and east of Swanborough Manor (merging with the Fourth Terrace). Further occurrences are near Northease Farm [4145 0640] (p. 80), west of Rodmell and south of Southease.

Superficially the Second and First terraces have a loamy consistency and only rarely could a significant gravel content be proved by augering (to 1.4 m). Kirkaldy and Bull (1940, p.133) described a 'low terrace at 46 ft [14 m]' OD at Sharpsbridge: 'The face of the pit shows 3 feet [0.9 m] of unstratified loam without flints resting on 5 feet [1.5 m] of ferruginous gravel ... The gravel is roughly stratified with somewhat coarser bands interposed by finer material, and shows a tendency to become coarser towards the bottom. The bulk of the gravel is composed of sandstone and ironstone from the Hastings Beds, flints are rare in the upper 2½ feet [0.8 m] but they increase in number downwards. The base is not seen for the pit floods with water, and it is only after pumping that the lowest layer can be examined. This contains large brown flints and rare pebbles of quartz and other erratic rocks ...' Two areas of First Terrace have been mapped near this location but the pit described is no longer extant.

Small spreads of First Terrace deposits border the river 1.5 km WSW of Uckfield. A temporary section at the Uckfield sewerage works [4637 2046] showed Upper Tunbridge Wells Sand overlain by 1.4 m of gravel, beneath up to 0.5 m of sandy Head. The gravel was composed of pebbles of sandstone and tabular ironstone not greater than 50 mm in diameter; no flints were found. Downstream [455 170], near Isfield, Kirkaldy and Bull (1940) described 'a pit at 40 ft [12 m] O.D.' showing 'in general ... two feet [0.6 m] of soil with scattered small flints passing down into 2½ feet [0.8 m] of roughly stratified sandy material with some pebbles of tabular sandstone, below which was a bed two to three feet [0.6 to 0.9 m] in thickness and made conspicuous by two to six inch [0.08 to 0.15 m] bands stained black or brown; the pebbles were mostly sandstone with a black coating, ironstone and towards the base flints. At the bottom and generally under water, was coarser material, consisting mainly of large mahogany flints, which had been angular, but whose edges were generally well rounded. Ironstone and sandstone pebbles were also common ... Very occasionally light coloured flints, quartzites and other erratic pebbles were to be found.' This section is located in Second Terrace deposits.

Extensive spreads of the lowest two terraces occur from Isfield southwards to South Malling. A river-bank section [4285 1355] near Wellingham exposed up to 1.2 m of brown to ochreous silty loam with a variable content of buff angular flints (Second Terrace). CRB,RDL

The terrace deposits at South Malling consist of brown silty and sandy loam. Flint pebbles are abundant in the soil and are accompanied by a few scattered small Wealden sandstone pebbles. A small section [4078 1152] 150 m NW of Old Malling Farm showed the following: Rubble consisting of subangular to angular blocks of flint up to 0.15 m in diameter in a chalky matrix; a few pipes in the chalky rubble infilled with brown silty loam containing numerous flints and fine-grained brown sandstone pebbles, 1.2 m, all overlain by brown silty loam up to 0.9 m thick (Second Terrace). The base of the terrace deposit here appeared to be piped into pre-existing chalky, flinty Head or Coombe Deposits. White (1926) gave a brief description of the 'Valley Gravel', composed of angular flints and a few flint pebbles, then exposed in the railway cutting [4124 1130] approximately 300 m ESE of Old Malling Farm. This cutting must formerly have exposed both Third Terrace and Second Terrace deposits, though only 'Valley Gravel' was distinguished here in the previous survey and it is impossible to say to which terrace White's description applies. Kirkaldy and Bull (1940) described an erosion flat at about 60 ft [18 m] OD on this spur, the surface being covered by sand overlain by 'flinty drift', and both being well exposed in the railway cutting in 1940; it is assumed that these deposits and the 60-foot erosion surface are part of what has been mapped as Second Terrace. BY

Further occurrences of Second Terrace and First Terrace were mapped between Iford and Rodmell. The deposit at Piddinghoe which is ascribed to the Second Terrace has been mentioned above (p. 82). White (1926, p.70) referred to an Acheulian implement found at Piddinghoe (?Second Terrace) and one of Chelles pattern from Rodmell (?Second-First Terrace). RDL

Cuckmere River valley

Small areas of First Terrace deposits have been mapped locally in the valleys of tributaries of the Cuckmere River near Lions Green and Horam [e.g. 5565 1850; 5575 1600; 5860 1735]. The deposits consist of brown silty loam containing scattered sandstone, siltstone and ironstone pebbles. BY

Much of the Second Terrace deposits around Hellingly and Horsebridge superficially has a loamy consistency (to 1.4 m depth). A river-bank section [5878 1282] near Shawpits Farm exposed 2.4 m of iron-stained loam with subordinate gravels of Wealden origin. Kirkaldy and Bull (1940, p.131) described a pit [586 124] at Hellingly 'showing four feet [1.2 m] of loam overlying with a sharp junction a gravel composed of smoothed pebbles of ironstone and sandstone derived from the Hastings Beds. This gravel is roughly bedded with some seams of sand and with many of the larger pebbles lying horizontally ... the gravel was proved to a thickness of 8½ feet [2.6 m]. It rested on two feet [0.6 m] of yellow clay containing a few pebbles, mostly large brown flints ... This gravel has yielded quartzite and other erratic pebbles. ...'

A temporary section at Lower Horsebridge [5781 1145] displayed blue-grey clay (Weald Clay) overlain by 1.5 m of loamy ferruginous

gravels, vaguely stratified (Second Terrace), and by 0.9 m of silty loam (Head). About 4.9 m of loam ascribed to the Second Terrace were proved in the Hailsham Borehole [5746 1083] overlying Weald Clay.

Festoons of gravels consisting of pebbles of Wealden rocks up to 0.6 m thick and resting on Weald Clay occur on the slope to the tributary stream near Goldup Shaw [571 096]; they occur at levels up to 18 m OD. Surface washes of brown-patinated flints commonly overlie the terrace loams in the vicinity of Michelham Priory [563 093].

Excavations within the core of the meander of the Cuckmere River for the clear-water tank site at Arlington Reservoir showed the following sections:

	Thickness m
1 [5397 0697]	
Silty clay, massive, grey and ochreous mottled (?Head)	2.4 to 3.0
Gravel, fine- to medium-grained, ferruginous, with sand partings and water-bearing clay lenses	up to 1.8
Weald Clay	—
2 [5394 0692]	
Sandy clay, pale grey and ochreous mottled; root traces; probably as below but oxidised by ground-water oscillation	2.4
Sandy clay, grey, with plant debris; leached to pale greenish grey at the top	1.5
Sandy clay, grey and green, with plant debris; finely laminated; flint pebbles locally near the base	1.2
Weald Clay	—

The gravel deposits in the first section were channelled into the sandy clays of the second, which were clearly waterlain in the lower part. The flank of the channel was oriented in a WNW–ESE direction, approximately at right angles to the general flow of the river. These deposits are therefore somewhat enigmatic, since they appear to represent fluviatile deposits, either in situ or reworked by solifluction in the upper part. Much of this drift material appears to have been removed or disturbed during the subsequent construction work: the area is shown as Weald Clay on the 1:50 000 geological map. RDL

A small area of brown silty loam, referred to the First Terrace of the Cuckmere River, has been mapped around and immediately south of Arlington village [5440 0740]. The deposit, which forms a poorly defined flat terrace feature approximately 2.5 m above the alluvium surface, consists of a brown, silty loam with abundant small pebbles of brown, fine-grained Wealden sandstone, siltstone and some ironstone, together with some brown angular flints.

On the west bank of the Cuckmere River 120 m west of the railway bridge, an old sand-pit [5342 0614] exposed 1 m of brown sandy loam with pockets up to 1 m across of angular to subangular flint pebbles (up to 5 cm in diameter) together with occasional small (up to 1.5 cm) dark brown fine-grained Wealden sandstone pebbles. This deposit lies roughly 6 m above the present river level. It may, in part, represent a terrace deposit, but it appears to form part of the blanket of Head which covers the ground to the north and west of the old pit and it is possible that terrace material has been incorporated in this solifluction material.

A small patch of dark brown silty loam containing abundant dark brown sandstone pebbles up to 4 cm in diameter stands as an island [5297 0562] about 1 m above alluvium level in the floodplain about 600 m NNW of Sherman Bridge. Similar First Terrace material covers a small part of the lower slopes on the west bank of the Cuckmere River [5274 0570] approximately 750 m NW of this bridge.

Brown silty loams of the Second and First terraces occupy a gently sloping bench on the east bank of the Cuckmere immediately south of the A27 road at Sherman Bridge. A small section [5318 0503] 20 m south of the bridge showed typical grey Gault clay overlain by 0.75 m of gravel composed of flattened ovoid pebbles of brown fine-grained sandstone and siltstone of Wealden origin and up to 2.5 cm in diameter, together with scattered subangular flints up to 7.5 cm in diameter, in a sparse, brown, sandy clay matrix. Very similar material in the First Terrace was seen in the sides of a farm track on the west bank of the river at Berwick Court [5258 0455]. Small sandstone pebbles, also assumed to be of Wealden origin, are common in the soil covering the sandy terrace deposits immediately south of here. A small area of Second Terrace and possibly Third Terrace has been recognised west and south-west of Berwick Court. Poorly developed terrace features are locally apparent and scattered Wealden sandstone pebbles are present in the soil. Brown sandstone pebbles up to 5 cm in diameter were noted in the spoil from a trench on the east side of the road [5240 0418] about 300 m SW of Berwick Court.

Milton Court [5265 0377] stands on a relatively flat First Terrace bench about 1 m above the modern alluvium. A small area of First Terrace has been mapped in Alfriston, surrounding the bare Chalk outcrop on which St Andrew's Church [5215 0300] stands. BY

ALLUVIUM AND PEAT

Surficially the alluvium in the valleys of the Lewes district consists of clayey silts and silty clays except in the Ashdown Beds catchment area where sandy loams are present. The Rivers Ouse and Cuckmere are sluggish streams with flood-plains of variable width. In each case, the alluvium fills a channel the base of which is graded to a lower sea level than that now prevailing, and consists of gravels, alluvial mud and silt. Near the mouth of the Ouse the base of the alluvial fill is at 29.6 m below OD.

White (1926, p.75) noted the following sequence of alluvial deposits near Lewes:

5 Soil and made ground
4 Brown flood loam with freshwater shells passing down to grey marly loam. Local peat lenses up to 1.5 m assumed to be meander cut-off accumulations
3 Blue-grey, stiff marly clay with sandy beds. Freshwater/estuarine shells in the upper part
2 Old soil?: horizon of wood, nuts etc.
1 Chalk and flint rubble or chalky marl
  Chalk bed-rock

After drilling a large number of boreholes within The Brooks south of Lewes (Plate 8), Jones (1971) showed the sequence in the main buried channel to be three-fold in nature:

3 Grey silt and silty clay, brown in the upper horizons, with shells. Local interbedded silts and sands. Up to 8 m thick.

Plate 8   The Brooks and Glynde Gap viewed from near Kingston (A 9947); inward-facing escarpments of the Kingston–Beddingham Anticline near Lewes

2 Interbedded peats, peaty clays and greenish grey clays. Peat development is greatest in the upper part. The lower peats contain considerable quantities of wood. Up to 10 m thick.

1 Sandy clay, sand and gravel up to 3.3 m thick.

Radio-carbon dates revealed ages extending back to 6290 ± 180 years BP (at 8.15 m below OD) for samples taken from within the peaty deposits (Jones, 1971, p.45). Pollen spectra obtained from these deposits showed high arboreal pollen values indicating that the Downs were well wooded in Mesolithic times (Thorley, 1971). Outside the main buried channel Jones found that the silts overlap the peaty deposits to rest on bed-rock. Local truncation of the lower peaty beds indicates a period of minor incision of the Ouse drainage. The full drift sequence rests on a surface with a minimum elevation of 13.1 m below OD north of Rise Farm.

In the Cuckmere valley, excavations for the Arlington Reservoir revealed that a buried channel has a basal surface that shows evidence of more than one phase of incision and that extends to 4.9 m below OD. Radio-carbon dating of

peaty material from the base of the alluvial fill gave an age of 9435 ± 120 BP (*Radiocarbon*, Vol. 13, 26–8).

Shallow BGS boreholes drilled in 1971 in the Pevensey Levels (Figure 28) proved the following typical sequence:

		Thickness m
4	Soft brown clay	1.5 to 2.1
3	Peat	0.6 to 1.8
2	Soft, pale to dark grey clay, rarely shelly, with local sand lenses	1.8 to 9.1 +
1	Firm ochreous clayey silt, shelly with fragments of peat, sandstone and flint	0.6 to 1.8

In or near former river-courses the peat bed (3) was found to be absent. Since some boreholes failed to reach bed-rock, bed 1 was used as a secondary datum level. Information from these and other boreholes indicates that the alluvium extends down to about 15 m below OD and probably has a complex basal surface. It is likely that deeper channels are present nearer the coast.

Sandy intercalations were proved near the present

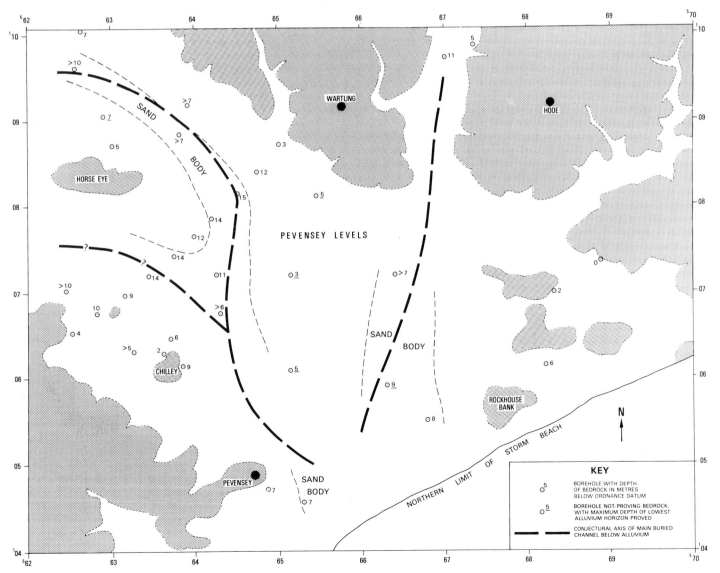

Figure 28   The alluvium of the central part of the Pevensey Levels

seaward margin of these beds and east of Horse Eye (Figure 28). In some cases these sands occur within the thick drift sequences above channels and may represent multi-layer channel-fill deposits. The possibility that some may represent barrier-bar deposits cannot be excluded, however, in view of the paucity of borehole information available. A basal peat horizon was proved in one borehole [about 6825 0615] at the Star Inn pumping station at 6.4 m below OD.

Mr M.J. Hughes (personal communication) has reported that the foraminifera collected from one borehole [6266 1001] north of New Bridge indicate a brackish-water environment with sand-flat associations initially, tending to a high-marsh environment in the upper levels of bed 2. A macrofauna including *Scrobicularia* and *Cardium* is abundant in the debris from ditch-cuts which penetrate bed 2. Large trunks of trees are found within the peat bed (3) (see, for example, Mantell, 1833, pp.19–20).

The landward boundary of the Pevensey Levels is generally marked by gentle slopes falling to alluvium level. Only at Rockhouse Bank [677 055] is there a steep slope on the south side, suggesting the presence of a former marine cliff, although artificial earthworks have modified the profile.

In the area of Willingdon Levels [610 035], a sequence of deposits similar to that of Pevensey Levels is apparently present. Boreholes indicate that the alluvium extends down to 15 m below OD. Jennings and Smyth (1982) have described the coastal deposits which occur immediately to the south of this district.
RDL

RECLAMATION OF THE PEVENSEY LEVELS

The history of the reclamation of the marshland at Pevensey has been detailed by Salzman (1910) and Dulley (1966). The following description is largely a summary of the latter account. It is probable that much of the reclamation work was carried out between 1086 and 1287. The marshland was protected from the sea by the bank of storm beach shingle, so that this work largely involved improvement of the internal drainage system and prevention of flooding back from the outlets to the sea south of Pevensey and near Rockhouse Bank (Figure 29). One of the centres from which reclamation proceeded was the ditched enclosure on Moat Marsh [661 060]. Other centres were probably at Manxey and Horse Eye, both of which supported communities at that time. The distribution of the major embankments is shown in Figure 29.

The former existence of medieval saltworks, which extracted brine from the mud of unreclaimed saltings, is shown by mounds of mud at a number of localities, particularly near Pevensey. The mounds are 'between three and five feet [0.9 to 1.5 m] high, irregularly oval in shape, and about fifty yards [45.7 m] in average diameter, although there is considerable variation in size' (Dulley, 1966, p.28). Most of the mounds are near former tidal channels and in some cases they were incorporated into the embankments.

Before the great flood of 1287, the tide 'flowed between the sea walls as far as Rickney Bridge and along the Old Haven from Pevensey to Waller's Haven, as well as in the Esthaven [also known as Godyngeshaven], while between Pevensey and the sea was an archipelago of islands of marsh or shingle intersected by tidal creeks' (Dulley, 1966, p.32). Following the inundation several attempts were made to improve the drainage system and to protect the area from flooding. The Commissioners of Sewers for the Sussex Coast were appointed in 1289 and they dammed the Pevensey Haven 'either at the modern Pevensey Bridge or, more probably, just to the north, at the junction with the Old Haven' (Dulley, 1966, p.32). In later years modifications were made to the outlet near Pevensey and new cuts were made between the Old Haven and Godyngeshaven (A-B followed by C-D, Figure 29). The cut C-D was extended (E-F) in 1455. Neglect of the sea walls in the Bestenover area after about 1540 caused this ground to revert to salt marsh and the Godyngeshaven to become ineffective. The channel south of Pevensey, which reached the sea at Wallsend, 'was affected by shingle drift, so that the mouth moved rapidly eastward, leaving a long, narrow channel parallel with the coast as far as the original Godyngeshaven mouth' (Dulley, 1966, p.34). After 1623 a new sluice was constructed near Wallsend across the horse-shoe bend and later, about 1694, a further sluice was constructed at the mouth of the haven and the remaining tidal creeks were drained. It was at this time that Pevensey ceased to be a port.
RDL

DETAILS

River Ouse valley

At Uckfield a site investigation borehole [4727 2091] proved 9.7 m of 'Alluvium peat and sand and gravel' above grey silt of the Lower Tunbridge Wells Sand.
CRB

Within the narrow channel of the River Ouse at the site of the bridge-crossing in Lewes [4193 1040], boreholes proved alluvium extending down to 10 m below OD. Peaty beds appear to infill the irregular upper surface of the Coombe Deposits. One borehole [4187 1034] (ground level 4.11 m OD) proved the sequence:

	Thickness m	Depth m
MADE GROUND	1.37	1.37
ALLUVIUM		
Soft fissured grey silty clay	3.66	5.03
Stiff brown silty clay with some fine gravel	1.37	6.40
COOMBE DEPOSITS		
Dense flint [sic]	0.46	6.86
Brown chalk cobbles with putty chalk	6.86	13.72
CHALK		
White chalk	—	—

Three peat samples from a trial borehole [426 093] at 3.13 m OD drilled for the Lewes by-pass site investigation gave the following radio-carbon dates:

Sample No.	Depth in metres	Years BP
EK 190	5.0 to 5.15	4346 ± 40
EK 191	5.15 to 5.3	4305 ± 40
EK 192	5.5 to 5.7	5000 ± 40

A site investigation borehole [4218 0917] (ground level 5.87 m OD) in The Brooks, north-east of the Upper Rise, showed the following sequence:

	Thickness m	Depth m
MADE GROUND	3.60	3.60
ALLUVIUM		
Soft grey and brown silty clay with peat	1.40	5.00
Soft grey, locally very silty, clay with occasional shell fragments	2.50	7.50

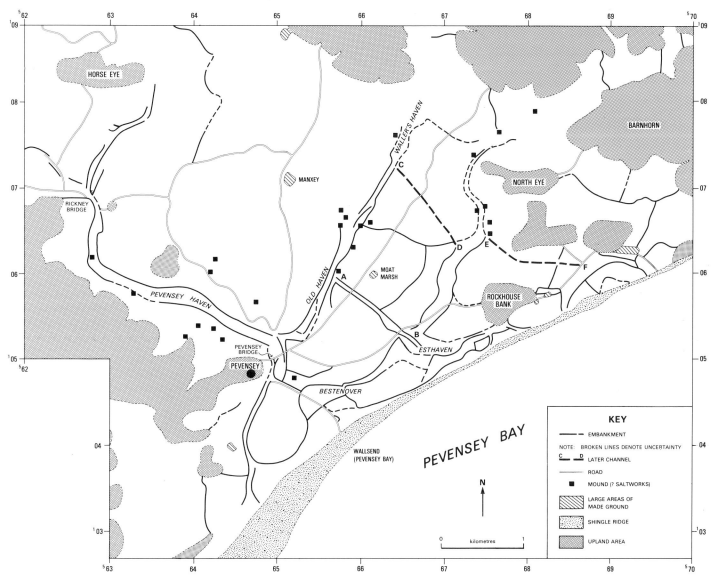

Figure 29   The reclamation of the Pevensey Levels (with some information from
Dulley, 1966)

	Thickness m	Depth m
Grey silty sand with fine organic material	3.20	10.70
Soft grey silty clay with fine organic material	1.05	11.75
Fibrous peat in clay matrix	1.25	15.00
Soft grey very silty clay with traces of peat	1.40	14.40
COOMBE DEPOSITS		
Grey chalk fragments in a grey silty clay matrix	1.50	15.90
LOWER CHALK		
Grey chalk	1.10	17.00
		RDL

Laughton Level

The recent survey has shown that the considerable area of alluvium
south-west of Laughton, at the head of Glynde Reach and known as
Laughton Level, is rather less extensive than was formerly mapped.
The thin alluvium present in the area appears to consist mostly of a
grey to greyish brown clay. Augering near the margins of the
alluvium suggests that a thin gravel occurs locally at its base,
though the composition of this gravel is unknown.

Whitaker and Reid (1899) gave the following details of a well at
the mill at Glynde [4578 0870] which passed through the alluvium
in the relatively narrow channel of Glynde Reach immediately
south-west of the alluvium basin:

	Thickness m
ALLUVIUM	
Mould	2 ft [0.6 m]
Clay, stone coloured, firm	2½ ft [0.7 m]
Black sand, with hazel nuts and sticks	11½ ft [3.5 m]
?HEAD — Chalk-rubble	3 ft [0.9 m]
LOWER CHALK — Chalk-rock	100 ft [30.5 m]

Alluvium down to a depth of 5.03 m was proved in a trial borehole
[4585 0865] drilled at Glynde Station in 1965. Mantell (1833)
recorded the discovery on the west side of Glynde Bridge of a paved
Roman causeway together with a brass coin of Antoninus Pius

'... lying three feet beneath the turf, upon a bed of silt twenty feet thick ...' No description was given of the material covering the causeway, though as White (1926) commented, if the material is alluvial in origin a significant increase in the thickness of the alluvium has occurred here since Roman times.

Cuckmere River valley

Narrow belts of alluvium floor many of the tributaries of the Cuckmere River extending in many cases to the heads of the valleys. In the majority of the tributary valleys the alluvial fill is relatively thin, consisting mainly of dark brown to greyish brown silty loam, locally with small sandstone, siltstone and a few ironstone pebbles.

In places rather broader stretches of alluvium have accumulated within historical times and these mark the sites of former hammer ponds. The remains of dams are still to be seen at many such locations.   BY

Pevensey Levels

The alluvium sequence has been described above (p.85). Figure 29 shows the present-day geography of the area, with former river courses, and enclosures, delineated by the relict embankments.

Peat deposits over 1.2 m thick occur at the surface in tributary valleys to the Pevensey Levels [6165 1105; 6220 1105] near Magham Down; in one case [6165 1105] the peat forms a raised mass.

In the northern area, Whelpley Level, a peat horizon occurs extensively at depths of 0.9 to 1.5 m. Exposures in ditch-cuts show that this body has irregular upper and lower surfaces displaying contorted and chevron profiles. *Cardium* is common in the clays adjacent to the peat bed.

The peat horizon is reported to be present at a greater depth (2.4 m) in the northern part of Horse Eye Level. The debris from ditch-cuts in this area shows organic clays and silts with subordinate peat fragments. At Rickney a peat bed was proved in boreholes [627 070] at 4 m depth. Fine sands were thrown from ditch excavations [6211 0846] immediately west of Horse Eye and around Newhouse [647 075]. An equivalent silt horizon was proved in a borehole [6384 0881] at 2.6 m depth and sand was penetrated in an auger hole [6479 0650] at 1 m depth near Manxey Sewer, although an intervening borehole [6433 0674] proved sands at 3.9 m depth.

A radio-carbon date of 3715 ± 80 years BP was obtained by Mr B. Moffat (personal communication) from a peat stratum at 2.8 m depth in the tributary valley [694 087] to the south of Hooe in the Hastings district.   RDL

STORM GRAVEL BEACH DEPOSITS

Storm Gravel Beach Deposits are composed of shingle piled up by wave action on the shore. They occur as a single ridge (or 'full') in the area east of Pevensey but south-westwards this feature merges with the shingle complex at the Crumbles in the Eastbourne district. Boreholes just to the east of the Lewes district [689 058] have shown that the shingle may be in excess of 4.5 m thick; locally the base of the deposits extends down to 4.5 m below OD.

The shingle full has acted as a protective feature for the Pevensey Levels to the north (p. 86). It has probably existed in some modified form since mid-Flandrian times (about 5000 years BP), when the post-glacial rise in sea level showed a marked decrease in rate.   RDL

MADE GROUND

Several areas of Made Ground have been delineated within the district. Spreads of domestic and industrial refuse cover parts of the alluvium in the Ouse valley between South Malling [410 110] and Lewes [420 100] and near the railway immediately south of Lewes [4220 0930]. Small areas of quarry refuse, composed largely of top soil and chalk debris, are found associated with most of the larger disused chalk-pits in the area.

In addition, miscellaneous deposits include the large earth-fill dam at Arlington [5360 0715], together with small associated areas of fill in the former channel of the Cuckmere River, and the smaller ancient earthwork at Burlough Castle [5303 0420] near Milton Street.   BY

SUPERFICIAL STRUCTURES

The outcrops of Wealden rocks have been affected locally by superficial movements under gravity. Landslips, on a small scale, have been recognised on the steeper valley sides within the clay formation outcrops.

Cambering and the associated valley-bulging, which are generally ascribed to periglacial activity, have been recognised in the district. Cambering occurs where competent sandstones (such as the Ardingly Sandstone) overlie thick argillaceous beds on the influves of the more deeply incised streams. Valley-bulges, which result from the movement of clay in the valley floor at the foot of cambered slopes, have been revealed by the steeply dipping and contorted strata encountered in stream sections. The presence of bedding-plane shears in the clay formations (p. 75) may have provided an agency whereby mass-movement took place. From regional knowledge it appears that valley-bulges may be present throughout the district, wherever valleys are floored with strata which were particularly susceptible to disturbance by permafrost action, including clay formations and the Chalk. These effects are particularly evident in the deep valleys floored with Purbeck Beds or Wadhurst Clay. However, a valley-bulge structure was noted in the Weald Clay of the Cuckmere valley (see below), which has a shallow profile. In the Purbeck Beds outcrop, disturbance of the surficial strata is probably a common feature but the scale of movement is not known because of the nature of the terrain and lack of exposure. The Broadoak Borehole [6195 2214] demonstrated the presence of open fracturing in the Purbeck strata down to approximately the level of the valley floor.   RDL

Landslips

In the area of the Purbeck inliers limited evidence for landslips was noted in areas of open farmland. It is probable that old landslips have been stabilised and are no longer recognisable in the extensive woodlands hereabouts. Slipped areas on Ashdown Beds were noted in a field [6215 2145] north of Bingletts Wood. Small landslips are present in the open ground [6250 2155 to 6280 2160] to the east and affect the Purbeck Beds and lowest Ashdown Beds. Hummocky ground and evidence for minor slips were noted along the valleyside eastwards to the Willingford Stream, particularly on the outcrop of the beds above the Greys Limestones. Small areas of slipped ground are present on the steep valley slopes near the Obelisk [670 212].

Some 1.7 km east of Little Horsted a 400-m strip of slipped Wadhurst Clay borders the east bank of a small stream [4876 1868]. Erosion of the east bank results in continuing small slips of this unstable ground.

A small area of landslipped Wadhurst Clay has been mapped on the east side of the stream approximately 800 m SW of Iwood Place Farm, Warbleton [6273 1615]. The slipping here has apparently resulted from the oversteepening of the bank of the stream.

Landslipped ground [633 138] is present on the Wadhurst Clay outcrop in the valley north of Chilsham. To the south of Stunts Green, landslips have affected the Wadhurst Clay below the spring line at the base of the Tunbridge Wells Sand [628 124].

South of Barcombe Cross hummocky sloping ground is present in two areas [416 154; 422 152] where terrace deposits overlie the Weald Clay. Superficial movements here were probably enhanced by the action of springs.

The steep south-east slope of Rockhouse Bank [679 056] shows evidence of minor landslip movements of the Weald Clay.   RABB,CRB,RDL,BY

Valley-bulges

In the Dudwell valley many small ephemeral exposures show evidence of valley-bulging. Contorted Purbeck mudstones were noted in Church Wood [6103 2160; 6117 2159]. Dip values for an exposure [6149 2167] in Church Wood indicated folding about an west–east axis. Sandy carbonaceous mudstone nearby [6155 2169] showed nearly vertical bedding with the same strike trend. Interbedded limestones and shales [6172 2183] in the tributary valley showed a variety of dip values from 0° to 90° and a west–east strike. Very contorted shales [6233 2177] were noted to the east of the lake near Tottingworth Wood. Contorted shales and limestones with generally vertical dip values were recorded at three adjacent localities downstream [6243 2181; 6247 2184; 6251 2185]. Vertical shales were also seen in the tributary valley [6239 2167] to the south. Vertical shales and sandy limestones with a west–east strike were noted below Milkhurst Wood [6252 2190]. An exposure

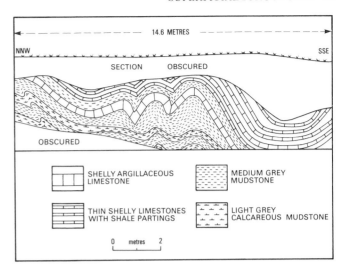

Figure 30 Superficial structures in the Purbeck Beds at Poundsford Ford

nearby [6250 2189] showed a section 10 m long in contorted beds of the same lithology. A little farther downstream [6281 2186] overturned beds of sandy limestones, sandstones and shales with plants were seen. Vertical shales with limestones were observed near Ten Acre Wood [6308 2193]. A sharp anticlinal structure was noted in Nine Acre Wood [6368 2187] affecting paper shales and porcellanous limestones. The bulge structure at Poundsford Ford [6385 2167] is shown in Figure 30. In the valley [5785 2010] south of Heathfield sands and silts in the Ashdown Beds are affected by valley-bulging. In the Ashdown Beds, vertical dips were recorded in Gameland Wood [6072 2142]. In the Willingford Stream valley, contorted sandstones and siltstones in the Ashdown Beds were noted at one locality [6521 2073] and vertical beds of the same nature, striking parallel to the valley sides (NNW–SSE), were recorded a little upstream [6528 2054].   RABB

In the stream [5856 1864] 390 m ENE of the village school at Maynard's Green, siltstones and clays overlain by siltstones and sandstones (Ashdown Beds) are folded into a small anticline, the axis of which strikes roughly north–south, parallel to the stream course.

Dips of up to 75°, almost certainly an effect of valley-

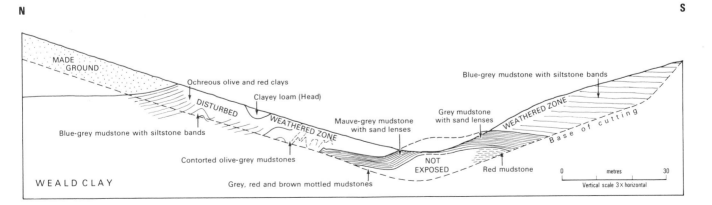

Figure 31 Section in excavations for the northern end of the Arlington Reservoir dam

bulging, occur in sandstones and clays in the Ashdown Beds in the stream [5963 1939] 500 m north of Nettlesworth Place. Valley-bulging is evident in the Ashdown Beds at Caller's Corner, north of Warbleton, where Ashdown Beds sandstone, siltstone and clays exposed in the east bank of the stream [6114 1950] dip upstream at up to 22°. Steep dips up to 75° SSE were seen in valley-bulged Ashdown Beds sandstones and siltstones in the Clippingham Stream [6264 1686] approximately 480 m east of Durrants Farm, near Rushlake Green.

Mudstones with thin beds of siltstone, belonging to the Wadhurst Clay, were seen in a small anticline, striking north-east, in a gully [5906 1784] 330 m west of the Brewers' Arms public house, Vine's Cross.   BY

Excavations in the Weald Clay made in 1969 for the north part of the Arlington Reservoir dam revealed extensive bulge-structures on the northern flank of the valley [538 075] (Figure 31). The severely contorted beds, which showed a chevron-style of folding, were restricted to a zone of some 25-m width. An isolated bulge-feature occurred 16 m farther up the valley slope. The amplitude of the flexures was generally of the order of 1 m. At one place (see Figure 31) it appeared that silty Head deposits were preserved in a synclinal structure as a result of the folding.   RDL

Collapse depressions

On the north side of the stream approximately 400 m SW of Court Horeham a circular hole [6072 1486] roughly 10 m in diameter and estimated to be about 4.5 m deep is probably a result of the collapse of the basal beds of the Wadhurst Clay into a cavity in the top Ashdown sandstones resulting from the enlargement of joints by the movement of underground water. Similar collapse features have been described from the Tunbridge Wells area by Bazley (*in* Bristow and Bazley, 1972).   BY

CHAPTER 10

Economic geology

Apart from the provision of water supply from underground sources the main local industries based on mineral products are the manufacture of bricks from clays in the Wealden Beds and of Portland Cement from the Chalk. Gypsum and small quantities of aggregates are obtained from one underground mine in the extreme north-eastern part of the district. Building stone and natural gas have also been products of minor consequence in the past and the area was formerly an important centre of the Wealden iron industry.

AGRICULTURE AND SOILS

The relatively heavy soils on the steep valley sides of the Purbeck Beds are mostly covered in woodland. The lighter, generally sandy and silty, poor soils covering the Ashdown Beds outcrop support a good deal of woodland and pastureland with relatively little arable cultivation. Local thin clay beds within the Ashdown Beds give rise to heavier, wetter soils, generally with conspicuous spring lines along valley sides. The Wadhurst Clay gives a heavy clay soil typically supporting pasture; areas of woodland are common, the woods near the base of the formation masking numerous traces of old bell-pits dug for ironstone. Much of the Tunbridge Wells Sand outcrop is marked by light, free-draining silty and sandy soils, like those of the Ashdown Beds. Woodlands are common but a larger area of the outcrop is covered by arable land. Soils developed on the Weald Clay in many respects resemble those of the Wadhurst Clay, though the soils on the Weald Clay are usually more silty. Red clay horizons are often conspicuous in the soil after ploughing and these give rise to poorly drained areas. Much of the Weald Clay outcrop is pasture covered, though a considerable area is under arable cultivation, especially where there are extensive thick spreads of loamy Head.

The outcrop of the Lower Greensand is reputed locally to support some of the best arable land in the area. Where the formation is not extensively drift-covered the sands, sandy clays and silts give a light loam. The brown silty Head which blankets much of the outcrop also gives rise to a light loamy soil well suited to cereal crops. Except where covered by superficial deposits the Gault outcrop is marked by very heavy clay soil, generally poorly drained and almost entirely used as pasture-land. Formerly much of the higher downland on the Chalk was rough grazing, but the majority of this ground is now under arable cultivation though the steep scarp slopes mostly remain as pasture. Small areas of typical Chalk grassland still exist in places, with patches of Clay-with-flints covered by gorse scrub. The alluvial ground of The Brooks, Laughton Level and Pevensey Levels is mostly meadow and pasture.   BY

WEALDEN IRON INDUSTRY

Iron ore was dug and smelted in the Weald in pre-Roman times. The industry was developed by the Romans and after a subsequent decline it again grew in importance in the Middle Ages, to reach a peak in Tudor times. The ironstone was dug from shallow shafts or 'bell-pits', or it was obtained as a by-product of the excavation of clay for marling from open pits. The latter method appears to have been more common in the present district. Straker (1931) gave a full account of the working, production and history of the industry and Shephard-Thorn and others (1966, pp.106–107) have provided a brief resumé. The last furnace in the Weald closed in 1826, although there were subsequent abortive attempts to revive the industry (Bristow and Bazley, 1972, p.114).

The greater part of the ore was obtained from the basal beds of the Wadhurst Clay; subordinate sources included ferruginous siltstone in the Hastings Beds, siderite beds in the Weald Clay and the Folkestone Beds (Sweeting, 1944, p.5). In the Lewes district, apart from the main Wadhurst Clay source, only limited occurrences of siderite in the middle Weald Clay were used for ore extraction. Straker (1931) recorded that in this district bloomeries were situated at Framfield, at Little Streele (Framfield), at Scallow Bridge (East Hoathly), at Little London (Waldron), at Cross in Hand, at Browndown (Heathfield), at Knowle Wood and Cindergill (Horam), at Marle Green, at Grove Hill (Hellingly), at Summertree Farm and Peartree (Warbleton), and at Herrings, Pagden Wood and Thorndale (Dallington). There were furnaces at Pounsley and New Place (Framfield), at Waldron, Woodman's (Warbleton), Heathfield, and Rushlake Green. Furnaces with forges were described at Tickerage (Framfield), Stream Furnace (Chiddingly), Battsford (Trulilow's Bridge) and Cowbeech. Other forges were recorded at Warbleton Priory and Steel Forge (Warbleton).

All of the above localities lie on the Hastings Beds outcrop. The area underlain by the Weald Clay was extensively dug in the past for marl and, to a lesser extent, for stone (Large-'*Paludina*' limestone for the most part). In some instances iron ore may have been obtained as a by-product. In the present survey, ironstone horizons in the Weald Clay, with associated large pits were noted near Town Littleworth, Whitesmith and Nash Street, and bell-pits were recorded at Dallas Lane [435 173].   RDL

FULLER'S EARTH

Two beds of fuller's earth 0.23 and 0.32 m thick, with Ca-montmorillonite contents of 90 per cent and 75 to 80 per cent respectively, were found near the middle of the Lower Greensand sequence in the Hampden Park Borehole [6120 0204] (Young, 1974; Young and others, 1978).   RJM

PHOSPHATIC CHALK

Strahan (1896) recorded a lenticular deposit of phosphatic chalk up to 2.74 m thick and at least 18.3 m wide in the old chalk-pit [near 4266 0942] at Southerham Works Quarry (see p. 72). A comparable deposit, which occurs in the *Terebratulina lata* Zone, was exposed in 1975 and was seen to comprise a phosphatic hardground and associated beds. Analyses revealed a P_2O_5 content of up to 5 per cent in phosphatic nodules extracted from the hardground; the chalk immediately beneath the hardground showed only 3 per cent P_2O_5. The deposit is too small and of too low a grade to be of economic interest at present.

GYPSUM

In the extreme north-east of the Lewes district gypsum is worked underground from the lower Purbeck Beds at Brightling Mine [6770 2187], the entrance of which is in the Tunbridge Wells district. Four gypsum seams varying from 1.4 to 4.3 m in thickness are present in the lowest 15.2 m of Purbeck Beds. At Brightling both the bottom seam (No.4) and the top seam (No.1) were worked until 1973, since when the poor quality of the No.4 seam has restricted mining to the No.1 seam. A 'pillar and room' mining method is followed, giving a rate of extraction of up to 75 per cent. The output is carried by a 5.6-km aerial ropeway to the Mountfield works [720 195] in the Hastings district, where most of the gypsum is used for wall-plaster and plaster-board manufacture. Brightling Mine is unique in being Britain's only 'safety light' gypsum mine; an explosion occurred in 1965 following a seepage of methane gas in the No.1 seam workings.

OIL AND NATURAL GAS

Topley (1875, p.38) recorded trials carried out on shales from the Purbeck Beds at Swife's Farm [6170 2282], north of Cade Street in the Tunbridge Wells district, where tar, pitch, naphtha and paraffin were extracted experimentally. No further work appears to have been carried out. A sand within the Tunbridge Wells Sand at Chilley, north-west of Pevensey, is known to be bituminous; Mantell (1833) described this sand as containing 15.4 per cent of bitumen. Minor oil seepages are known from the present gypsum workings in the Purbeck Beds at Brightling Mine and similar oil seepages were noted in fresh core specimens from this level in the Broadoak Borehole [6195 2214].

Gas was first discovered in the area in 1893, during the boring of a well for water in the stable yard of the New Heathfield Hotel [5815 2143]. A subsequent borehole [5804 2135] drilled for water in the nearby railway cutting in 1896 struck gas in the Hastings Beds at a depth of about 95 m, the gas-flow being sufficient to give a flame 4.9 m high when ignited. As boring proceeded the flow of gas increased, the

borehole being terminated at a depth of 114.9 m (Dawson, 1898). This gas was used to light Heathfield Station [5810 2130] until the early 1960's. An analysis of the gas made in 1902 is as follows (White, 1926):

	per cent
Carbon monoxide	1.00
Methane	93.16
Ethane	2.94
Nitrogen or other inert gas (by difference)	2.90
Total	**100.00**

Following the discovery of gas at Heathfield a number of boreholes were put down in the area but although gas was found to be widely present in the Ashdown Beds it was nowhere in commercial quantities. Both this gas and the oil occurrences mentioned previously probably have their source at depth in the bituminous shales of the Kimmeridge Clay.

A number of exploratory boreholes have been sunk on anticlinal folds within the district at Hellingly and Pevensey in search of oil or gas, but without success. Small quantities of methane ('marsh gas') were encountered in a number of shallow boreholes drilled in the alluvium of Pevensey Levels in 1971.

LIGNITE

Topley (1875, p.347) recorded small-scale trials, in 1801, of lignite from the Ashdown Beds in Waldron Gill, between Waldron and Heathfield, though the amount obtained here was probably very small. Lignite was also recorded by Topley (1875, p.348) from Brown's Lane, Waldron, again in the Ashdown Beds, and Mantell (1833, p.206) mentioned a bed of lignite up to 0.28 m thick in the Tunbridge Wells Sand at Newick Old Park. None of these sites can be precisely located at the present time.

BUILDING STONE

Most of the harder rocks of the district have been employed locally as building stones, though they are little used today. Sandstones from the Hastings Beds may be seen in a number of buildings in the north of the area on the Hastings Beds outcrop; Possingworth Manor House [5349 2049] near Blackboys, the remains of Holy Trinity Priory [6408 1812] east of Rushlake Green, and the walls enclosing Heathfield and Brightling parks [590 210; 680 206] provide good examples. Similar sandstones from the Tunbridge Wells Sand have also found a local use. Mantell (1833) recorded the occurrence of blocks of Tunbridge Wells sandstone, quarried at Chilley, in the Roman walls of Pevensey Castle.

The Large-'*Paludina*' limestone beds in the Weald Clay of the Laughton area, known locally as 'Sussex Marble' or 'Laughton Stone', have provided the building stone for a number of barns and farmyard walls in the neighbourhood (Plate 4). Good examples may be seen in the farm buildings at Home Farm, Laughton [4974 1307]. Individual limestone beds are up to 0.23 m thick. According to Topley (1875, p.108) considerable quantities of this limestone were dug at Laughton and flooded and degraded pits can still be seen in the area.

Glauconitic sandstones in the walls of Beddingham church [4450 0790] and Pevensey Castle [6445 0480] are said by White (1926, p.84) to have been obtained from the Upper Greensand at Eastbourne.

Throughout much of the southern and south-western part of the district flint has been much used as a building stone. Large beach cobbles of flint form an important part of the walls of Pevensey Castle, where they are accompanied by much smaller quantities of clay-ironstone nodules derived from the Wadhurst Clay and Weald Clay. Many buildings, especially at Lewes, Alfriston and Pevensey, also incorporate flint beach cobbles. Unworn flint has been much used locally for rough stone work, for example in the walls of St Pancras Priory, Lewes [4143 0954], where flint is accompanied by 'Sussex Marble', Bembridge Limestone, oolitic limestone and Caen Stone (White, 1926, p.84). Dressed flint facings are seen in Lewes Old Grammar School [4115 1000], St Michael-in-Lewes Church, Lewes [4132 1000], All Saints Church, Laughton [5007 1258], and St Andrews Church, Alfriston [5215 0300].

BRICK CLAY

Bricks, tiles and pottery have, in the past, been produced from a variety of clays within the district. Two brickworks are at present operating, at Hamsey [3985 1600] and Chailey [3935 1760], both working Weald Clay.

Weathered silts, clayey silts in the Ashdown Beds, and possibly some overlying Pleistocene Head, have been worked [6353 2005] for brickmaking near Three Cups Corner, Dallington. The local name 'Crock Kiln Wood' [5740 1990] applied to a small area of woodland to the west of Waldron Gill near Runtington Manor Farm suggests that clay beds present here in the Ashdown Beds may have been worked in the past.

Bricks were produced from the Wadhurst Clay at the Marle Green Brickworks [5905 1595] near Horam until 1972. A subsequent investigation of the site has been made with a view to opening a large new works capable of handling up to 573 cubic metres of clay per week. However, at the time of writing plans to develop this work have been shelved.

The small brickworks [4780 1417] working clays and silty clays in the Weald Clay on the north side of the A265 road approximately 2.5 km NE of Ringmer closed in 1971. The Cuckmere brickworks at Berwick Station [5260 0678] closed in 1957. Here bricks were made from a mixture of Weald Clay and glauconitic loamy sand from the lower part of the Lower Greensand. A number of old brickworks, now abandoned, worked the Weald Clay in the Hailsham and Polegate areas (p. 35), and earthenware was formerly manufactured from Weald Clay at the Dicker Potteries [5681 1114].

In the past bricks were made from the Gault at Ringmer, Norlington, Laughton, Ripe and Polegate, and locally drift deposits have been used, such as Pleistocene loams at Alfriston and recent floodplain loam at Piddinghoe (White, 1926, p.85).

NATURAL AGGREGATE

Waste rock, consisting largely of limestone, shale and anhydrite, from the Brightling gypsum mine has a widespread local use for hardcore and making up farm tracks. The waste products from the mine are separated at the Mountfield works [720 195] and are sold together with similar waste from Mountfield Mine.

Old pits [5702 1901] in a hard sandstone in the Ashdown Beds near Little London are said locally to have provided a source of aggregate for the construction of the railway through Horam. Topley (1875, p.59) referred to 'A hard, ferruginous, gritty sandstone … dug for roads in the woods near Little London'.

A number of small pits have provided small quantities of sand for local use from both the Ashdown Beds and the Tunbridge Wells Sand. Greater quantities were dug from the sand at the top of the Lower Greensand, the local equivalent of the Folkestone Beds, in a number of large pits around Ripe and Chalvington.

Chalk is used locally as a fill material for making roads and for strengthening river banks.

Gravels were formerly worked from the terraces of the River Ouse around Barkham, Isfield and Barcombe (p. 82). The gravels consist of a mixture of sandstone and ironstone pebbles derived from the Hastings Beds, together with flints, quartz and other rock-types (Kirkaldy and Bull, 1940) (see also p. 78). Similar terrace gravels in the Cuckmere valley have been worked at Horsebridge.

Considerable areas of beach shingle, composed almost entirely of flint, fringe the coast of Pevensey Bay and are worked a little way to the south of the present district.

LIME AND CEMENT

Limestones in the Purbeck Beds were formerly quarried and mined and either used in the production of ground limestone or burnt for lime. According to White (1926) the lime obtained from the 'Blues' limestone was preferred for building and that from the 'Greys' for agricultural lime.

The only limestone now quarried in the district is chalk and considerable quantities of this rock have been obtained from numerous pits in the past. A good deal of chalk is quarried for use as fill material. Much of the output from the Ash Chalk Pits [5640 0285], in the Lower Chalk near Wannock, is used for this purpose. Lime is still made by burning Lower Chalk at Glynde Station [4590 0855].

Portland Cement is produced at Lewes cement works, Southerham [4260 0950], where marly Lower Chalk from two nearby pits [4280 0900; 4320 0905] is blended to provide a feed for the kiln; no separate argillaceous material is added to the feed. The former Rodmell works [4355 0635], south of Beddingham, obtained chalk from the Lower Chalk and Middle Chalk. Clay was supplied from the top of the Gault in an adjacent pit [4400 0720]. In the past cement was produced from the Middle Chalk and Upper Chalk obtained from the large and now abandoned pits at Southerham [4255 0955; 4260 0965]; clay is believed to have been obtained from the old clay-pit [4690 0980] 1.5 km NE of Glynde Station. Cement was also produced from Upper Chalk at a works situated in the large abandoned quarry [4493 0328] between Tarring Neville and South Heighton.   BY

HYDROGEOLOGY AND WATER SUPPLY

This district lies within Hydrometric Areas 40 and 41, and the water resources are administered by the Southern Water Authority. Most of the water resources within the district are the responsibility of two private water companies, the Eastbourne Waterworks Company and the Mid-Sussex Water Company.

Early references to the water supply of the district include Whitaker and Reid (1899), Whitaker (1911), White (1926) and Edmunds (1928). During the war years, a well catalogue with hydrogeological notes was published (Buchan and others, 1940), and a more recent well catalogue was prepared by Cole and others (1965). The water resources were detailed in reports by the Sussex River Authority (1970) and by the Kent River Authority (1970). The Southern Water Authority has provided more recent data on abstraction and resources, including a report on water resources under Section 24 of the Water Act 1973, some of which were reproduced in Monkhouse and Richards (1983). The district also appears on the Hydrogeological Map for the South Downs published on a scale of 1:100 000 by the Institute of Geological Sciences and the Southern Water Authority in 1978.

The mean annual rainfall for the period 1941–70 varies from more than 900 mm on the high ground of the South Downs to about 700 mm on the low ground of the Pevensey Levels. The mean annual evaporation varies from about 370 to 490 mm. The district includes a major portion of the catchments of the rivers Ouse and Cuckmere. A number of small streams drain the Pevensey Levels directly to the sea in the south-east.

Approximately 17 million cubic metres per annum (m³/a) of water are abstracted annually from intakes in the rivers Ouse and Cuckmere with pumped storage reservoirs at Arlington [535 074] and Barcombe [441 150]. Groundwater abstraction is less than that of surface water, amounting to approximately 8.03 million m³/a, of which some 7.4 million m³/a is taken from the Chalk, 30 thousand m³/a from the Tunbridge Wells Sand, and 600 thousand m³/a from the Ashdown Beds. A negligible amount is abstracted from the Lower Greensand. Typical chemical analyses of groundwater from the district are shown in Table 2.

The *Ashdown Beds* constitute an important aquifer, but with silt and clay layers commonly present, the permeability, particularly in the vertical direction, tends to be low. Yields tend to be better in the upper part of the formation where the argillaceous beds are less common. On the outcrop, a borehole of 150 mm diameter penetrating 15 m of saturated sands may be expected on average to yield 1.4 litres per second (l/sec) for a drawdown of 5 m. Carefully designed sand screens and filter packs are necessary to avoid siltation problems. Slightly greater yields have been obtained from boreholes penetrating the Ashdown Beds where they are confined beneath the Wadhurst Clay. Larger yields, of the order of 5 to 10 l/sec, have been obtained from excavated shafts supplemented by horizontal adits; however, even this

form of construction has failed to obtain useful yields in locations where the silt and clay beds are well developed. In places, small springs are thrown out by silt and clay layers, but the flows are only sufficient to support small, local demands.

The groundwater from the outcrop of the Ashdown Beds typically has a total hardness (as calcium carbonate) in the range 50 to 150 milligrammes per litre (mg/l). When confined by the Wadhurst Clay, the total hardness is typically less than 120 mg/l. The chloride ion concentration is generally of the order of 30 mg/l. Iron is commonly present in concentrations of more then 5.0 mg/l; concentrations of as much as 30 mg/l have been recorded.

Although the *Wadhurst Clay* is generally impermeable, small yields of the order of 0.3 l/sec have been obtained from lenses of limestone and sandstone. However, the groundwater quality is usually poor, with total hardness values of up to 400 mg/l and high concentrations of iron.

The variations in lithology and preserved thickness of the *Tunbridge Wells Sand* limit its value as an aquifer. Faulting commonly divides the formation into discrete hydrogeological units that are too small to be viable for abstraction. In the west of the district, around Chailey and Uckfield, the aquifer is divided into an upper and a lower division by the Grinstead Clay. The mean annual infiltration is probably of the order of 150 to 180 mm. The mean yield of a borehole of 150 mm diameter penetrating 15 m of saturated sand is of the order of 1.5 l/sec for a drawdown of 5 m. An exceptional yield of 23.7 l/sec was obtained for a drawdown of 6.8 m from a 250 mm diameter borehole at the former Adam's Hole pumping station [614 113]; an unconfined thickness of 27 m of Tunbridge Wells Sand was penetrated. The groundwater is usually soft with a total hardness of 60 to 80 mg/l. The chloride ion concentration is typically about 40 mg/l. Rather harder water is apparently present in aquifers close to the Grinstead Clay.

The *Weald Clay* contains thin limestone and sandstone beds from which small supplies of 0.3 l/sec or less have been taken. The natural recharge of these beds is very limited and yields often fail with time as the storage is exhausted. The groundwater is usually of poor quality, being ferruginous and mineralised; a well at Upper Dicker [549 096] yielded water with 2000 mg/l of total dissolved solids.

East of the Cuckmere River, the *Lower Greensand* is too thin to be useful as an aquifer. To the west, the upper few metres are generally of fine-grained sand, but in the strata beneath, the proportion of argillaceous material lessens the permeability appreciably. In the outcrop area, there are a number of small-diameter boreholes typically yielding some 0.3 l/sec. Boreholes sunk in the Lower Greensand confined beneath the Gault near Chalvington [522 094], Selmeston [510 070] and West Firle [471 071] have provided better yields of the order of 0.5 to 0.8 l/sec, with overflowing conditions sometimes present. The groundwater of the Lower Greensand typically has a total hardness of less than 200 mg/l, although levels as high as 300 mg/l have been recorded. The water is commonly ferruginous with iron concentrations in excess of 0.5 mg/l.

The *Gault* yields small quantities of groundwater from the more sandy basal beds and from the silty horizons near the top. The water is sometimes mineralised and unpotable.

Table 2   Representative analyses of groundwaters from aquifers in the Lewes area

Location	Cow Wish Bottom	Offham	Selmeston	Iford
National Grid Reference	[4507 0405]	[4025 1205]	[5214 0800]	[4089 0716]
Date of analysis	Summer 1982	Summer 1982	September 1945	21 October 1957
Classification and	Drift 1.5	Drift ⎤	Lower Greensand ⎤	Drift 7.6 ⎤
thickness of strata (m)	Upper Chalk 74.6	Lower Chalk ⎦ 5.2	?Weald Clay ⎦ 51.8	Lower Chalk ⎦
				Gault 91.4
				Lower Greensand 3.6
Electrical conductivity (μmho/cm)	494	454	—	750
Total dissolved solids (mg/l)	—	7.5	714	489
pH	7.5	—	—	—
Calcium (mg/l)	98	105	—	—
Magnesium (mg/l)	3.1	2.2	—	—
Sodium (mg/l)	22	11	—	—
Potassium (mg/l)	2.8	2.3	—	—
Silica (mg/l)	—	5.2	—	—
Total iron (mg/l)	0.01	0.02	1.4	—
Total manganese (mg/l)	0.01	0.01	—	—
Chloride (mg/l)	39	21	174	98
Fluoride (mg/l)	0.08	0.07	—	—
Sulphate (mg/l)	—	—	—	—
Nitrogen (as N) (mg/l)	8.6	3.5	absent	—
Total hardness (as $CaCO_3$) (mg/l)	214	218	309	12
Alkalinity (as $CaCO_3$) (mg/l)	171	200	—	230

Location	Amberstone	East Hoathly	Cowbeech	Hempstead
National Grid Reference	[5983 1138]	[4949 1681]	[6101 1498]	[482 217]
Date of analysis	22 April 1970	March 1946	14 June 1982	1985
Classification and	Drift 6.6	?Upper Tunbridge	Drift 3.7	Wadhurst Clay
thickness of strata (m)	Tunbridge Wells	Wells Sand 12.8	Ashdown Beds 102.7	Ashdown Beds
	Sand 44.3	Grinstead Clay 5.2		
		Lower Tunbridge		
		Wells Sand 9.4		
Electrical conductivity (μmho/cm)	330	—	365	347
Total dissolved solids (mg/l)	258	228	—	—
pH	6.1	6.1	7.06	6.9
Calcium (mg/l)	7	—	4.8	12
Magnesium (mg/l)	27	—	1.2	2.4
Sodium (mg/l)	52	—	78	72
Potassium (mg/l)	—	—	0.2	1.4
Silica (mg/l)	9.0	—	13.16	16.8
Total iron (mg/l)	20.8	9.8	2.7	12
Total manganese (mg/l)	3.6	present	0.27	0.28
Chloride (mg/l)	46	23	26.4	28
Fluoride (mg/l)	—	—	—	—
Sulphate (mg/l)	34	—	24	24
Nitrogen (as N) (mg/l)	—	absent	0.38	0.1
Total hardness (as $CaCO_3$) (mg/l)	—	131	37	40
Alkalinity (as $CaCO_3$) (mg/l)	62	—	149	148

In the *Chalk*, groundwater flow and storage is predominantly through and within fissures. In the Upper and Middle Chalk, transmissivities determined from borehole pumping tests vary from less than 1000 square metres per day (m²/d) to more than 2500 m²/d. The more marly Lower Chalk is less permeable, and transmissivities of more than 200 m²d are unlikely. Natural groundwater levels may fluctuate annually through a range of 30 m. Borehole yields tend to be rather variable and are dependent upon the intersection of water-bearing fissures. A borehole of 610 mm diameter and 76 m depth at Cow Wish Bottom [451 041] yielded 47.4 l/sec for 2.7 m drawdown, while another of 200 mm diameter and 168 m depth at Saxon Down [437 099] gave 19.8 l/sec for 8.8 m drawdown. Large-diameter shafts may be capable of yields of more than 50 l/sec, especially where they have been deepened by boreholes. Yields from the Lower Chalk are usually less than 1.0 l/sec.

Newly completed boreholes are usually treated with hydrochloric acid to remove the drilling slurry. Such acid treatment may increase the yield of a borehole by 100 per cent or more above the untreated rate.

The quality of groundwater from the Chalk is generally good with the total hardness usually in the range 200 to 400 mg/l, and the chloride ion concentration less than 50 mg/l. Iron is generally absent. However, excessive pumping of wells in the coastal area may induce saline intrusion with a consequent rise in the chloride ion concentration.

The *superficial deposits* of the district are not significant as sources of groundwater. In the valley of the River Ouse, small yields have been obtained from sands and gravels within the alluvium, but the underlying aquifers tend to be more productive and more reliable. In the extensive alluvial deposits of the Pevensey Levels, the groundwater is commonly saline.   RAM

REFERENCES

ALLEN, P. 1941. A Wealden soil bed with *Equisetites lyelli* (Mantell). *Proc. Geol. Assoc.*, Vol. 52, 362–374.

— 1947. Notes on Wealden fossil soil-beds. *Proc. Geol. Assoc.*, Vol. 57, 303–314.

— 1949. Wealden petrology: The Top Ashdown Pebble Bed and the Top Ashdown Sandstone. *Q. J. Geol. Soc. London*, Vol. 104, 257–321.

— 1954. Geology and geography of the London–North Sea uplands in Wealden times. *Geol. Mag.*, Vol. 91, 498–508.

— 1959. The Wealden environment: Anglo-Paris Basin. *Philos. Trans. R. Soc. London*, Vol. 242, 283–346.

— 1960a. Geology of the Central Weald: a study of the Hastings Beds. *Geol. Assoc. Centen. Guide*, No. 24.

— 1960b. Strand-line pebbles in the mid-Hastings Beds and the geology of the London uplands. General features, Jurassic pebbles. *Proc. Geol. Assoc.*, Vol. 71, 156–168.

— 1961. Strand-line pebbles in the mid-Hastings Beds and the geology of the London uplands. Carboniferous pebbles. *Proc. Geol. Assoc.*, Vol. 72, 271–286.

— 1962. The Hastings Beds deltas: Recent progress and Easter Field Meeting Report. *Proc. Geol. Assoc.*, Vol. 73, 219–243.

— 1967a. Origin of the Hastings Facies in North-Western Europe. *Proc. Geol. Assoc.*, Vol. 78, 27–105.

— 1967b. Strand-line pebbles in the mid-Hastings Beds and the geology of the London uplands—Old Red Sandstone, New Red Sandstone and other pebbles. Conclusion. *Proc. Geol. Assoc.*, Vol. 78, 241–276.

— 1975. Wealden of the Weald: a new model. *Proc. Geol. Assoc.*, Vol. 86, 389–438.

— 1981. Pursuit of Wealden models. *J. Geol. Soc. London*, Vol. 138, 375–405.

AMEDRO, F., DAMOTTE, R., MANIVIT, H., ROBASZYNSKI, F. and SORNAY, J. 1978. Echelles biostratigraphiques dans le Cénomanien du Boulonnais. *Géol. Meditérr.*, Vol. 5, 5–18.

ANDERSON, F. W. 1940. Ostracod zones of the Wealden and Purbeck. *Adv. Sci.*, Vol. 1, 259.

— 1962. Correlation of the Upper Purbeck Beds of England with the German Wealden. *Liverpool, Manchester Geol. J.*, Vol. 3, 21–32.

— 1985. Ostracod faunas in the Purbeck and Wealden of England. *J. micropalaeontology*, Vol. 4, 1–68.

— and BAZLEY, R. A. B. 1971. The Purbeck Beds of the Weald (England). *Bull. Geol. Surv. G.B.*, No. 34.

— — and SHEPHARD-THORN, E. R. 1967. The sedimentary and faunal sequence of the Wadhurst Clay (Wealden) in boreholes at Wadhurst Park, Sussex. *Bull. Geol. Surv. G.B.*, No. 27, 171–235.

BAILEY, H. W., GALE, A. S., MORTIMORE, R. N., SWIECICKI, A. and WOOD, C. J. 1983. The Coniacian–Maastrichtian Stages of the United Kingdom, with particular reference to southern England. *Newsl. Stratigr.*, Vol. 12, 19–42.

— 1984. Biostratigraphical criteria for the recognition of the Coniacian to Maastrichtian Stage boundaries in the Chalk of north-west Europe, with particular reference to southern England. *Bull. Geol. Soc. Denmark*, Vol. 33, 31–39.

BARROIS, C. 1876. Recherches sur le Terrain Crétacé Supérieur de l'Angleterre et de l'Irlande. *Mém. Soc. Géol. Nord.*, Vol. 1.

BÖHM, J. 1915. Zusammenstellung der Inoceramen der Kreideformation (Nachtrage). *Jahrb. Preuss. Geol. Landesanst.*, Vol. 35, 595–599.

BRISTOW, C. R. and BAZLEY, R. A. B. 1972. Geology of the country around Royal Tunbridge Wells. *Mem. Geol. Surv. G.B.*

BROMLEY, R. G. 1967. Some observations on burrows of thalassinoidean Crustacea in chalk hardgrounds. *Q. J. Geol. Soc. London*, Vol. 123, 157–182.

— 1978. Hardground diagenesis. In *Encyclopedia of Earth Sciences, VI. The Encyclopedia of Sedimentology.* FAIRBRIDGE, R. W. and BOURGEOIS, J. (Editors). (Stroudsburg.)

— and EKDALE, A. A. 1984. Trace fossil preservation in flint in the European Chalk. *J. Paleontol.*, Vol. 58, 298–311.

— and GALE, A. S. 1982. The lithostratigraphy of the English Chalk Rock. *Cretaceous Res.*, Vol. 3, 273–306.

— SCHULZ, M-G. and PEAKE, N. B. 1975. Paramoudras: giant flints, long burrows and early diagenesis of chalks. *K. Dansk. Vidensk. Selsk. Geol. Skr.*, 20/10.

BRYDONE, R. M. 1917. The base of the Chalk Zone of *Holaster planus* in the Isle of Wight. *Geol. Mag.*, Vol. 4, 245–249.

BUCHAN, S., ROBBIE, J. A., HOLMES, S. C. A., EARP, J. R., BUNT, E. F. and MORRIS, L. S. O. 1940. Water supply of S.E. England from underground sources. *Wartime Pam.*, No. 10, Part IX.

BULL, A. J. 1936. Studies in the geomorphology of the South Downs (Eastbourne to Arun Gap). *Proc. Geol. Assoc.*, Vol. 47, 99–129.

BUTLER, D. E. 1981. Marine faunas from concealed Devonian rocks of southern England and their reflection of the Frasnian transgression. *Geol. Mag.*, Vol. 118, 679–697.

CALLOMON, J. H. 1968. The Kellaways Beds and the Oxford Clay. Chapter 14. Pp.264–290 in *The Geology of the East Midlands.* SYLVESTER BRADLEY, P. C. and FORD, T. D. (Editors). (Leicester University Press.)

— and COPE, J. C. W. 1971. The stratigraphy and ammonite succession of the Oxford and Kimmeridge clays in the Warlingham Borehole. *Bull. Geol. Surv. G.B.*, No. 36, 147–176.

CARTER, D. J. and HART, M. B. 1977. Aspects of mid-Cretaceous stratigraphical micropalaeontology. *Bull. Br. Mus. Nat. Hist. (Geol.)*, Vol. 29, 1–135.

CASEY, R. 1950. The junction of the Gault and Lower Greensand in East Sussex and at Folkestone, Kent. *Proc. Geol. Assoc.*, Vol. 61, 268–298.

— 1954. *Falciferella*, a new genus of Gault ammonites, with a review of the family Aconeceratidae in the British Cretaceous. *Proc. Geol. Assoc.*, Vol. 65, 262–277.

— 1960a. 'Hengestites', a new genus of Gault ammonites. *Palaeontol.*, Vol. 2, 200–209, Pl. 29.

— 1960b. A monograph of the Ammonoidea of the Lower Greensand, Part I. *Palaeontogr. Soc. (Monogr.)*, i–xxxvi, 1–44.

— 1961.   The stratigraphical palaeontology of the Lower Greensand.   *Palaeontol.*, Vol. 3, 487–621.

— 1963.   The dawn of the Cretaceous Period in Britain.   *Bull. South-east Union Sci. Soc.*, No. 117, 15 pp.

— 1978, 1980.   A Monograph of the Ammonoidea of the Lower Greensand, parts VIII and IX.   *Palaeontogr. Soc. (Monogr.)*

CHAPMAN, F. 1896.   Appendix on the foraminifera and ostracoda in STRAHAN, 1896.   *Q. J. Geol. Soc. London*, Vol. 52, 470–473.

COLE, M. J., MATTHEWS, A. M., ROBERTSON, A. S. and others. 1965.   Records of wells in the area of New Series One-Inch (Geological) Lewes (319), Hastings (320) and Dungeness (321) sheets.   *Nat. Environ. Res. Counc.*

COX, B. M. and GALLOIS, R. W. 1981.   The stratigraphy of the Kimmeridge Clay of the Dorset type area and its correlation with some other Kimmeridgian sequences.   *Rep. Inst. Geol. Sci.*, No. 80/4.

DAVISON, C. 1924.   *A History of British Earthquakes.* (Cambridge.)

DAWSON, C. 1898.   On the discovery of natural gas in East Sussex.   *Q. J. Geol. Soc. London*, Vol. 54, 564–571.

— and WOODWARD, A. S. 1913.   On the discovery of a Palaeolithic skull and mandible in a flint-bearing gravel overlying the Wealden (Hastings Beds) at Piltdown, Fletching (Sussex).   *Q. J. Geol. Soc. London*, Vol. 69, 117–144.

— — 1914.   Supplementary note on the discovery of a Palaeolithic human skull and mandible at Piltdown, Sussex.   *Q. J. Geol. Soc. London*, Vol. 70, 82–93.

DESTOMBES, J. P. and SHEPHARD-THORN, E. R. 1971.   Geological results of the Channel Tunnel site investigation 1964–65.   *Rep. Inst. Geol. Sci.*, No. 71/11.

DINES, H. G., BUCHAN, S., HOLMES, S. C. A. and BRISTOW, C. R. 1969.   Geology of the country around Sevenoaks and Tonbridge.   *Mem. Geol. Surv. G.B.*

DIXON, F. 1850.   *The geology and fossils of the Tertiary and Cretaceous formations of Sussex.*   1st Edition.

DREW, F. 1861.   On the succession of beds in the 'Hastings Sand' in the northern portion of the Wealden area.   *Q. J Geol. Soc. London*, Vol. 17, 271–286.

DRUMMOND, P. V. O. 1985.   The *Micraster* biostratigraphy of the Senonian White Chalk of Sussex, southern England.   *Géol. Meditérr.*, Vol. 10, 177–182.

DUFF, K. L. 1978.   Bivalvia from the English Lower Oxford Clay (Middle Jurassic).   *Palaeontogr. Soc. (Monogr.).*

DULLEY, A. J. F. 1966.   The level and port of Pevensey in the Middle Ages.   *Sussex Archaeol. Collect.*, Vol. 104, 26–45.

EDMUNDS, F. H. 1928.   Wells and springs of Sussex.   *Mem. Geol. Surv. G.B.*

EKDALE, A. A. and BROMLEY, R. G. 1984.   Comparative ichnology of shelf-sea and deep-sea chalk.   *J. Paleontol.*, Vol. 58, 322–332.

ELLIS, C. S. 1985.   Chalk dry valley fills in East Sussex.   *Brighton Polytech. Geogr. Soc. Mag.*, No. 11, 35–42.

ELSDEN, J. V. 1887.   Superficial geology of the southern portion of the Weald area.   *Q. J. Geol. Soc. London*, Vol. 43, 637–656.

ERNST, G. 1971.   Biometrische Untersuchungen über die Ontogenie und Phylogenie der *Offaster/Galeola*-Stammesreihe (Echin.) aus der nordwest-europäischen Oberkreide.   *N. Jb. Geol. Paläont. Abh.*, Vol. 139, 169–225.

FALCON, N. L. and KENT, P. E. 1960.   Geological results of petroleum exploration in Britain 1945–1957.   *Mem. Geol. Soc. London*, No. 2, 56 pp.

GALE, A. S. and SMITH, A. B. 1982.   The palaeobiology of the Cretaceous irregular echinoids *Infulaster* and *Hagenowia*.   *Palaeontol.*, Vol. 25, 11–42.

— and WOODROOF, P. B. 1981.   A Coniacian ammonite from the 'Top Rock' in the Chalk of Kent.   *Geol. Mag.*, Vol. 118, 557–560.

GALLOIS, R. W. and COX, B. M. 1976.   The stratigraphy of the Lower Kimmeridge Clay of eastern England.   *Proc. Yorkshire Geol. Soc.*, Vol. 41, 13–26.

— and MORTER, A. A. 1982.   The stratigraphy of the Gault of East Anglia.   *Proc. Geol. Assoc.*, Vol. 93, 351–368.

GASTER, C. T. A. 1920.   An undescribed species of *Trochiliopora*.   *Geol. Mag.*, Vol. 57, 526.

— 1929.   Chalk zones in the neighbourhood of Shoreham, Brighton and Newhaven, Sussex.   *Proc. Geol. Assoc.*, Vol. 39, 328–340.

— 1937a.   The stratigraphy of the Chalk of Sussex, Part I: The West Central area—Arun gap to valley of the Adur, with zonal map.   *Proc. Geol. Assoc.*, Vol. 48, 356–373.

— 1937b.   Field meeting at Lewes, Sussex.   *Proc. Geol. Assoc.*, Vol. 48, 354–355.

— 1939.   Stratigraphy of the Chalk of Sussex, Part II: Eastern area—Seaford to Cuckmere valley and Eastbourne, with zonal map.   *Proc. Geol. Assoc.*, Vol. 50, 510–526.

— 1951.   The stratigraphy of the Chalk of Sussex, Part IV: East Central area—Between the valley of the Adur and Seaford with zonal map.   *Proc. Geol. Assoc.*, Vol. 62, 31–64.

GODWIN-AUSTEN, R. 1851.   On the gravel beds of the valley of the Wey.   *Q. J. Geol. Soc. London*, Vol. 7, 278–288.

GRAMBAST, L. 1974.   Phylogeny of the Charophyta.   *Taxon*, Vol. 23(4), 463–481.

HANCOCK, J. M. 1975.   The petrology of the Chalk.   *Proc. Geol. Assoc.*, Vol. 86, 499–535.

— and KENNEDY, W. J. 1967.   Photographs of hard and soft chalks taken with a scanning electron microscope.   *Proc. Geol. Soc. London*, No. 1643, 249–252.

HARRIS, C. S. 1982.   *Albian microbiostratigraphy (foraminifera and ostracoda) of south-east England and adjacent areas.*   D.Phil. Council National Academic Awards, Plymouth Polytechnic.

HART, M. B. 1973.   A correlation of the macrofaunal and microfaunal zonations of the Gault Clay in southeast England.   Pp. 267–288 in *The Boreal Lower Cretaceous*. CASEY, R. and RAWSON, P. F. (Editors).   *Spec. Issue Geol. J.*, No. 5. (Liverpool: Seel House Press.)

HOFKER, J. 1959.   Les foraminiferes de Crétacé supérieur de Harmignies, Basin de Mons.   *Ann. Soc. Géol. Belg.*, Vol. 82, 319–333.

HOLLIDAY, D. W. and SHEPHARD-THORN, E. R. 1974.   Basal Purbeck evaporites of the Fairlight Borehole, Sussex.   *Rep. Inst. Geol. Sci.*, No. 74/4.

HOWITT, F. 1964.   Stratigraphy and structure of the Purbeck inliers at Sussex (England).   *Q. J. Geol. Soc. London*, Vol. 120, 77–113.

JEANS, C. V., MERRIMAN, R. J., MITCHELL, J. G. and BLAND, D. J. 1982.   Volcanic clays in the Cretaceous of Southern England and Northern Ireland.   *Clay Minerals*, Vol. 17, 105–156.

JEFFERIES, R. P. S. 1961.   The palaeoecology of the *Actinocamax plenus* Subzone (Lowest Turonian) in the Anglo-Paris Basin.   *Palaeontol.*, Vol. 4, 609–647.

— 1963. The stratigraphy of the *Actinocamax plenus* Subzone (Turonian) in the Anglo-Paris Basin. *Proc. Geol. Assoc.*, Vol. 74, 1–33.

JENNINGS, S. and SMYTH, C. 1982. A preliminary interpretation of coastal deposits from East Sussex. *Quat. Newsl.*, No. 37, 12–19.

JOHNSON, J. P. 1901. Some sections in the Cretaceous rocks around Glynde, and their fossil contents. *Geol. Mag.*, Vol. 8, 249–251.

JONES, D. K. C. 1971. The Vale of the Brooks. *Inst. Brit. Geogr. Conf.*, 43–46.

JUKES-BROWNE, A. J. and HILL, W. 1900. The Cretaceous Rocks of Britain. Vol. 1. The Gault and Upper Greensand of England. *Mem. Geol. Surv. G.B.*

— 1903. The Cretaceous Rocks of Britain, Vol. 2. The Lower and Middle Chalk of England. *Mem. Geol. Surv. G.B.*

— 1904. The Cretaceous Rocks of Britain, Vol. 3. The Upper Chalk of England. *Mem. Geol. Surv. G.B.*

KAYE, T. 1966. Lower Cretaceous palaeogeography of north-west Europe. *Geol. Mag.*, Vol. 103, 257–262.

KENNEDY, W. J. 1967. Field meeting at Eastbourne, Sussex, Lower Chalk sedimentation. *Proc. Geol. Assoc.*, Vol. 77, 365–370.

— 1969. The correlation of the Lower Chalk of south-east England. *Proc. Geol. Assoc.*, Vol. 80, 459–560.

— and JUIGNET, P. 1974. Carbonate banks and slump beds in the Upper Cretaceous (Upper Turonian–Santonian) of Haute Normandie, France. *Sedimentol.*, Vol. 21, 1–42.

KENT RIVER AUTHORITY. 1970. First periodical survey of water resources. (Maidstone: Kent River Authority.) 108pp.

KIRKALDY, J. F. 1935. The base of the Gault in Sussex. *Q. J. Geol. Soc. London*, Vol. 91, 519–537.

— 1937. The overstep of the Sandgate Beds in the Eastern Weald. *Q. J. Geol. Soc. London*, Vol. 93, 94–126.

— and BULL, A. J. 1936. Field meeting at Berwick and Hellingly, Sussex. *Proc. Geol. Assoc.*, Vol. 47, 265–268.

— 1940. The geomorphology of the rivers of the southern Weald. *Proc. Geol. Assoc.*, Vol. 51, 115–150.

LAKE, R. D. 1975a. The stratigraphy of the Cooden Borehole near Bexhill, Sussex. *Rep. Inst. Geol. Sci.*, No. 75/12.

— 1975b. The structure of the Weald—a review. *Proc. Geol. Assoc.*, Vol. 86, 549–557.

— and HOLLIDAY, D. W. 1978. Purbeck Beds of the Broadoak Borehole, Sussex. *Rep. Inst. Geol. Sci.*, No. 78/3.

— and SHEPHARD-THORN, E. R. In preparation. The geology of the country around Hastings and Dungeness. *Mem. Br. Geol. Surv.*

— and THURRELL, R. G. 1974. The sedimentary sequence of the Wealden Beds in boreholes near Cuckfield, Sussex. *Rep. Inst. Geol. Sci.*, No. 74/2.

— and YOUNG, B. 1978. Boreholes in the Wealden Beds of the Hailsham area, Sussex. *Rep. Inst. Geol. Sci.*, No. 78/23.

LAMPLUGH, G. W. 1920. The structure of the Weald and analogous tracts. *Q. J. Geol. Soc. London*, Vol. 93, 156–194.

LEES, G. M. and COX, P. T. 1937. The geological basis of the present search for oil in Great Britain by the D'Arcy Exploration Co., Ltd. *Q. J. Geol. Soc. London*, Vol. 93, 156–194.

LOMBARD, A. 1956. *Géologie sédimentaire: les séries marines.* (Paris and Liege.)

MANTELL, G. A. 1818. A sketch of the geological structure of the south-eastern part of Sussex. *Gleaners Portfolio, or Provincial Magazine, Lewes.*

— 1822. *The fossils of the South Downs: or illustrations of the geology of Sussex.* (London.)

— 1827. *Illustrations of the geology of Sussex, containing a general view of the south-eastern part of England.* (London.)

— 1833. *The geology of the south-east of England.* (London.)

— 1846. Description of some fossil fruits from the Chalk formation of the south-east of England. *Q. J. Geol. Soc. London*, Vol. 2, 51–54.

MONKHOUSE, R. A. and RICHARD, H. J. 1983. European Community Atlas of Groundwater Resources—United Kingdom. Environment and Consumer Protection Service, European Economic Community. (Hanover: Th. Schaeffer.)

MORTER, A. A. 1984. Purbeck–Wealden mollusca and their relationship to ostracod biostratigraphy, stratigraphical correlation and palaeoecology in the Weald and adjacent areas. *Proc. Geol. Assoc.*, Vol. 95, 217–234.

— and WOOD, C. J. 1983. The biostratigraphy of Upper Albian–Lower Cenomanian *Aucellina* in Europe. *Zitteliana*, Vol. 10, 515–529.

MORTIMER, M. G. and CHALONER, W. G. 1972. The palynology of the concealed Devonian rocks of Southern England. *Bull. Geol. Surv. G.B.*, No. 39.

MORTIMORE, R. N. 1979. *The relationship of stratigraphy and tectonofacies to the physical properties of the White Chalk of Sussex.* Unpublished PhD thesis, Brighton Polytechnic.

— 1983. The stratigraphy and sedimentation of the Turonian–Campanian in the Southern Province of England. *Zitteliana*, Vol. 10, 27–41.

— 1986a. Stratigraphy of the Upper Cretaceous White Chalk of Sussex. *Proc. Geol. Assoc.*, Vol. 97, 97–139.

— 1986b. Controls on Upper Cretaceous sedimentation in the South Downs with particular reference to flint distribution. In *The scientific study of flint and chert: papers from the Fourth International Flint Symposium*, Vol. 1. SIEVEKING, G. DE G. and HART, M. B. (Editors.) (Cambridge: Cambridge University Press.)

— and WOOD, C. J. 1986. The distribution of flint in the English Chalk, with particular reference to the 'Brandon Flint Series' and the high Turonian flint maximum. In *The scientific study of flint and chert: papers from the Fourth International Flint Symposium*, Vol. 1. SIEVEKING, G. DE G. and HART, M. B. (Editors.) (Cambridge: Cambridge University Press.)

NORRIS, G. 1969. Miospores from the Purbeck Beds and marine Upper Jurassic of southern England. *Palaeontol.*, Vol. 12, 574–620.

OAKLEY, K. P. 1955. The composition of the Piltdown hominoid remains. *Bull. Br. Mus. Nat. Hist. (Geol.)*, Vol. 2, 254–261.

OWEN, H. G. 1971. Middle Albian stratigraphy in the Anglo-Paris Basin. *Bull. Br. Mus. Nat. Hist. (Geol.) Suppl.*, Vol. 8, 1–164.

— 1976. The stratigraphy of the Gault and Upper Greensand of the Weald. *Proc. Geol. Assoc.*, Vol. 86 (for 1975), 475–498.

— 1979. Ammonite zonal stratigraphy in the Albian of North Germany and its setting in the Hoplitinid Faunal Province. *Aspekte der Kreide Europas*, IUGS Ser. A.6, 563–588.

— 1984. The Albian Stage: European Province chronology and ammonite zonation. *Cretaceous Res.*, Vol. 5, 329–344.

PENN, I. E., MERRIMAN, R. J. and WYATT, R. J. 1979. The Bathonian strata of the Bath–Frome area. *Rep. Inst. Geol. Sci.*, No. 78/22.

PENNING, W. H. and JUKES-BROWNE, A. J. 1881. The geology of the neighbourhood of Cambridge. *Mem. Geol. Surv. G.B.*

PEPPER, D. M. 1973. A comparison of the 'Argile à silex' of northern France with the 'Clay-with-flints' of southern England. *Proc. Geol. Assoc.*, Vol. 84, 331–352.

PHILIPS, W. 1819. *A selection of facts from the best authorities arranged so as to form an outline of the Geology of England and Wales.* (London.)

PRICE, F. G. H. 1874. On the Gault of Folkestone. *Q. J. Geol. Soc. London*, Vol. 30, 342–366.

— 1879. *A monograph of the Gault.* (London.) 81 pp.

PRICE, R. J. 1976. Palaeoenvironmental interpretations of the Albian of western and southern Europe, as shown by the distribution of selected foraminifera. *Maritime Sediments Spec. Publ.*, 1, Int. ser. part B, *Biostratigraphy and palaeoecology*, 625–648.

— 1977. The stratigraphical zonation of the Albian sediments of north-west Europe, as based on foraminifera. *Proc. Geol. Assoc.*, Vol. 88, 65–91.

RAWSON, P. F., CURRY, D., DILLEY, F. C., HANCOCK, J. M., KENNEDY, W. J., NEALE, J. W., WOOD, C. J. and WORSSAM, B. C. 1978. A correlation of Cretaceous rocks in the British Isles. *Spec. Rep. Geol. Soc. London*, No. 9.

REEVES, J. W. 1949. Surface problems in the search for oil in Sussex. *Proc. Geol. Assoc.*, Vol. 59 (for 1948), 234–269.

— 1958. Subdivision of the Weald Clay in Sussex. *Proc. Geol. Assoc.*, Vol. 69, 1–16.

ROWE, A. W. 1899. An analysis of the genus *Micraster*. *Q. J. Geol. Soc. London*, Vol. 55, 494–547.

— 1900. The zones of the White Chalk of the English Coast. 1—Kent and Sussex. *Proc. Geol. Assoc.*, Vol. 16, 289–368.

— 1908. The zones of the White Chalk of the English Coast. V—The Isle of Wight. *Proc. Geol. Assoc.*, Vol. 20, 209–352.

SALZMANN, L. F. 1910. The inning of Pevensey Levels. *Sussex Archaeol. Collect.*, Vol. 53, 30–60.

SHEPHARD-THORN, E. R., SMART, J. G. O., BISSON, G. and EDMONDS, E. A. 1966. Geology of the country around Tenterden. *Mem. Geol. Surv. G.B.*

— LAKE, R. D. and ATITULLAH, E. A. 1972. Basement control of structures in the Mesozoic rocks in the Strait of Dover region. *Philos. Trans. R. Soc. London*, Vol. A. 272, 99–113.

SHEPHERD, W. 1972. *Flint. Its origin, properties and uses.* (London.)

SMART, J. G. O., BISSON, G. and WORSSAM, B. C. 1966. Geology of the country around Canterbury and Folkestone. *Mem. Geol. Surv. G.B.*

SORNAY, J. 1978. Précisions paléontologiques et stratigraphiques sur divers Inocérames Cénomaniens, et, en particulier, sur ceux de la Sarthe figurés par E. Guéranger en 1867. *Geobios*, Vol. 11, 505–515.

SOUTHERN WATER AUTHORITY. 1980. Water Act 1973, Section 24(1A): Survey of existing water use and management. (Worthing: Southern Water Authority.) 25 maps.

SPATH, L. F. 1923. On the ammonite zones of the Gault and contiguous deposits. *Summ. Prog. Geol. Surv. G.B.* (for 1922), 139–149.

— 1930. On some Ammonoidea from the Lower Greensand. *Annu. Mag. Nat. Hist.*, Ser. (10), 5, 417–464.

— 1934. A monograph of the ammonoidea of the Gault. *Palaeontogr. Soc. (Monogr.)*, Part 9.

STEWART, D. J. 1983. Possible suspended-load channel deposits from the Wealden Group (Lower Cretaceous) of Southern England. *Spec. Publ. Int. Assoc. Sedimentol.*, Vol. 6, 369–384.

STOKES, R. B. 1975. Royaumes et provinces fauniques du Crétacé établis sur la base d'une étude systématique du genre *Micraster*. *Mém. Mus. Nat. Hist. Nat., Paris*, (c), Vol. 31.

STOPES, M. 1915. *Catalogue of the Mesozoic plants in the British Museum, (Natural History). The Cretaceous flora*, Part 2, *The Lower Greensand (Aptian) plants of Britain.* (London.)

STRAHAN, A. 1896. On a phosphatic chalk with *Holaster planus* at Lewes. *Q. J. Geol. Soc. London*, Vol. 52, 463–470.

STRAKER, E. 1931. *Wealden iron.* (London.)

SUSSEX RIVER AUTHORITY. 1970. Water for Sussex: First periodic survey of demand for water. (Brighton: Sussex River Authority.) 243pp.

SWEETING, G. S. 1944. Wealden iron ore and the history of its industry. *Proc. Geol. Assoc.*, Vol. 55, 1–20.

SYKES, R. M. and CALLOMON, J. H. 1979. The *Amoeboceras* zonation of the Boreal Upper Oxfordian. *Palaeontol.*, Vol. 22, 889–903.

TAITT, A. H. and KENT, P. E. 1958. Deep boreholes at Portsdown (Hants) and Henfield (Sussex). *Tech. Publ. British Petroleum.* (London: British Petroleum.)

THORLEY, A. T. 1971. Vegetational history in the Vale of the Brooks. *Inst. Br. Geogr. Conf.*, 47–50.

THURRELL, R. G., WORSSAM, B. C. and EDMONDS, E. A. 1968. Geology of the country around Haslemere. *Mem. Geol. Surv. G.B.*

TOPLEY, W. 1875. The geology of the Weald. *Mem. Geol. Surv. G.B.*

VOIGT, E. 1959. Die ökologische Bedeutung der Hartgründe ('Hardgrounds') in der oberen Kreide. *Paläont. Z.*, Vol. 33, 129–147.

— and HÄNTZSCHEL, W. 1956. Die grauen Bänder in der Schreibkreide Nordwestdeutschlands und ihre Deutung als Lebensspuren. *Mitt. Geol. Staatsinst. Hamburg*, Vol. 25, 104–122.

WHITAKER, W. 1908. The water supply of Kent from underground sources. *Mem. Geol. Surv. G.B.*

— 1911. The water supply of Sussex from underground sources (supplement). *Mem. Geol. Surv. G.B.*

— and REID, C. 1899. The water supply of Sussex from underground sources. *Mem. Geol. Surv. G.B.*

WHITE, H. J. O. 1924. Geology of the country near Brighton and Worthing. *Mem. Geol. Surv. G.B.*

— 1926. The geology of the country near Lewes. *Mem. Geol. Surv. G.B.*

— 1928. The geology of the country near Ramsgate and Dover. *Mem. Geol. Surv. G.B.*

WILKINSON, I. P. and MORTER, A. A. 1981. The biostratigraphical zonation of the East Anglian Gault by ostracoda. Pp.161–176 in *Microfossils from Recent and fossil shelf seas.* NEALE, J. W. and BRASIER, M. D. (Editors). (Ellis Horwood, Chichester and British Micropalaeontological Society.)

WILLIAMS, R. B. G. 1971. Aspects of the geomorphology of the South Downs. *Inst. Br. Geogr. Conf.*, 35–42.

WIMBLEDON, W. A. and COPE, J. C. W. 1978. The ammonite zonation of the English Portland Beds and the zones of the Portlandian Stage. *J. Geol. Soc. London*, Vol. 135, 183–190.

— and HUNT, C. O. 1983. The Portland–Purbeck junction (Portlandian–Berriasian) in the Weald, and correlation of latest Jurassic–early Cretaceous rocks in Southern England. *Geol. Mag.*, Vol. 120, 267–280.

WOOD, C. J., ERNST, G. and RASEMANN, G. 1984. The Turonian–Coniacian stage boundary in Lower Saxony (Germany) and adjacent areas: the Salzgitter-Salder Quarry as a proposed international standard section. *Bull. Geol. Soc. Denmark*, Vol. 33, 225–238.

WOODS, H. 1912. A Monograph of the Cretaceous Lamellibranchia of England, Part 8. *Palaeontogr. Soc. (Monogr.)*, Vol. 2, 285–340.

WOOLDRIDGE, S. W. and LINTON, D. L. 1955. *Structure, surface and drainage in south-east England*. (London.)

WORSSAM, B. C. 1973. A new look at river capture and at the denudation history of the Weald. *Rep. Inst. Geol. Sci.*, No. 73/17.

— and IVIMEY-COOK, H. C. 1971. The stratigraphy of the Geological Survey borehole at Warlingham, Surrey. *Bull. Geol. Surv. G.B.*, No. 36

— and TAYLOR, J. H. 1969. Geology of the country around Cambridge. *Mem. Geol. Surv. G.B.*

WRIGHT, C. W. and KENNEDY, W. J. 1981. The Ammonoidea of the Plenus Marls and the Middle Chalk. *Palaeontogr. Soc. (Monogr.)*

YOUNG, B. 1974. *Annual Report for 1973*. (London: Institute of Geological Sciences.)

— and LAKE, R. D. In press. Geology of the country around Brighton and Worthing. *Mem. Br. Geol. Surv.*

— and MONKHOUSE, R. A. 1980. The geology and hydrogeology of the Lower Greensand of the Sompting borehole, West Sussex. *Proc. Geol. Assoc.*, Vol. 91, 307–314.

— MORGAN, D. J. and HIGHLEY, D. E. 1978. New fuller's earth occurrences in the Lower Greensand of south-eastern England. *Trans. Inst. Min. Metall.*, Vol. 87, B93–B96.

APPENDIX 1

List of six-inch maps

The following is a list of the six-inch geological maps included, wholly or in part, in the 1:50 000 Geological Sheet 319, with the names of the surveying officers and the dates of the survey for each six-inch map. The surveyors are: Dr R. A. B. Bazley, Dr C. R. Bristow, Mr E. A. Edmonds, Dr R. W. Gallois, Mr R. D. Lake and Mr B. Young.

TQ 30 NE	Kingston near Lewes Lake	1973
TQ 30 SE	Rottingdean Lake	1973
TQ 31 NE	Chailey Bristow and Lake	1971–72
TQ 31 SE	Plumpton Lake and Young	1971–73
TQ 32 SE	Scaynes Hill Gallois	1964
TQ 40 NW	Rodmell Lake and Young	1973
TQ 40 NE	West Firle Young	1972–73
TQ 40 SW	Newhaven Lake	1973
TQ 40 SE	Denton Young	1973
TQ 41 NW	Barcombe Cross Bristow and Lake	1971, 1974
TQ 41 NE	Little Horsted Bristow	1970
TQ 41 SW	Lewes Lake and Young	1973
TQ 41 SE	Ringmer Young	1972
TQ 42 SW	Fletching Bristow and Gallois	1964
TQ 42 SE	Maresfield Bristow	1964
TQ 50 NW	Berwick Young	1972–73
TQ 50 NE	Hailsham Lake	1971, 1973
TQ 50 SW	Alfriston Young	1973
TQ 50 SE	Willingdon Lake	1973
TQ 51 NW	East Hoathly Young	1971
TQ 51 NE	Horam Young	1971
TQ 51 SW	Chiddingly Lake	1971
TQ 51 SE	Hellingly Lake	1971
TQ 52 SW	Hadlow Down Bazley	1964
TQ 52 SE	Heathfield Bazley	1964
TQ 60 NW	Hankham Lake	1968, 1971
TQ 60 NE	Wartling Lake	1968
TQ 60 SW	Westham Lake	1968, 1973
TQ 60 SE	Pevensey Bay Lake	1973
TQ 61 NW	Rushlake Green Young	1971–72
TQ 61 NE	Dallington Bazley	1965
TQ 61 SW	Herstmonceux Lake	1970
TQ 61 SE	Boreham Street Lake	1968
TQ 62 SW	Burwash Common Bazley	1963
TQ 62 SE	Brightling Bazley and Edmonds	1960, 1964

APPENDIX 2

Borehole records

The abridged logs of BGS cored boreholes are given below. A few of the holes lie just outside the boundaries of the Lewes district. The National Grid reference of each site is given, together with the six-inch registration number of the borehole in the Survey records at Keyworth.

TUNBRIDGE WELLS (303) SHEET

Broadoak Borehole
[TQ 6195 2214]; (TQ 62 SW 4); surface level 126.8 m above OD. Date 1974

	Thickness m	Depth m
CRETACEOUS		
Purbeck Beds (part) (base of Cinder Bed Member at 47.44 m)	47.44	47.44
JURASSIC		
Purbeck Beds (part)	86.23	133.67
Portland Beds	seen to 9.71	143.38

LEWES (319) SHEET

West Firle Borehole
[TQ 4676 0809]; (TQ 40 NE 2); surface level about 18 m above OD. Date 1973

	Thickness m	Depth m
QUATERNARY		
Head	4.00	4.00
CRETACEOUS		
Lower Chalk	14.12	18.12
Gault	seen to 1.23	19.35

Glyndebourne Borehole
[TQ 4420 1141]; (TQ 41 SW 16); surface level about 115 m above OD. Date 1973

	Thickness m	Depth m
QUATERNARY		
Hillwash	2.00	2.00
CRETACEOUS		
Lower Chalk	46.36	48.36
Gault	104.24	152.60
Lower Greensand	seen to 18.31	170.91

Alciston Borehole
[TQ 5045 0553]; (TQ 50 NW 78); surface level 50.3 m above OD. Date 1973

	Thickness m	Depth m
CRETACEOUS		
Lower Chalk	15.54	15.54
Gault	seen to 0.92	16.46

East Hoathly Borehole
[TQ 5186 1603]; (TQ 51 NW 1); surface level about 55 m above OD. Date 1974

	Thickness m	Depth m
CRETACEOUS		
Tunbridge Wells Sand	31.39	31.39
Wadhurst Clay	49.70	81.09
Ashdown Beds	seen to 22.36	103.45

Ripe Borehole
[TQ 5059 1052]; (TQ 51 SW 1); surface level about 17 m above OD. Date 1972

	Thickness m	Depth m
QUATERNARY		
Head	1.50	1.50
CRETACEOUS		
Lower Greensand	46.29	47.79
Weald Clay	158.86	206.65
Tunbridge Wells Sand	seen to 7.37	214.02

Hailsham Borehole
[TQ 5746 1083]; (TQ 51 SE 1); surface level 16.5 m above OD. Date 1972

	Thickness m	Depth m
QUATERNARY		
Terrace loam	4.90	4.90
CRETACEOUS		
Weald Clay	85.73	90.63
Tunbridge Wells Sand	seen to 39.52	130.15

Glynleigh Borehole
[TQ 6085 0637]; (TQ 60 NW 14); surface level about 3.7 m above OD. Date 1972

	Thickness m	Depth m
CRETACEOUS		
Tunbridge Wells Sand	41.00	41.00
Wadhurst Clay	30.40	71.40
Ashdown Beds	seen to 25.12	96.52

EASTBOURNE (334) SHEET

Rodmill Borehole
[TQ 6008 0070]; (TQ 60 SW 5); surface level 13.7 m above OD. Date 1973

	Thickness m	Depth m
CRETACEOUS		
Lower Chalk	33.10	33.10
Gault	seen to 0.50	33.60

Hampden Park Borehole
[TQ 6120 0204]; (TQ 60 SW 6); surface level 1.6 m above OD. Date 1973

	Thickness m	Depth m
QUATERNARY		
Alluvium	6.00	6.00
CRETACEOUS		
Gault	12.82	18.82
Lower Greensand	20.75	39.57
Weald Clay	seen to 50.53	90.10

APPENDIX 3

The Gault of the Glyndebourne Borehole

The broad lithological sequence proved in the Glyndebourne Borehole [4420 1141] and the basis of the zonal subdivision have been discussed on pp. 41 and 42. The following account provides more detailed information on the Gault that occurred between 48.35 and 152.60 m depth.

Stoliczkaia dispar Zone (upper part)
Depth 48.35 to 57.51 m

These beds are characterised by silts and calcareous silty mudstones, with a limited shelly fauna dominated by the bivalves *Aucellina gryphaeoides*, *Entolium orbiculare*, *Neithea (Neithella) sp.*, *Plicatula radiola gurgitis*, *Pycnodonte (Phygraea) vesicularis*; exogyrine oysters also occur, together with the scaphopod *Dentalium ?divisiense*. A specimen of *Lepthoplites falcoides* from 52.06 m is usually an indicator for the *rostratum* Subzone (see Owen, 1976).

The base of the upper part of the zone coincides with a slight lithological change at 57.51 m. Microfaunal evidence indicates that the *perinflatum* Subzone may be present above 50 m, represented by Zone Fii of Harris (1982) and Zone 8 of Price (1977), equivalent to the highest part of Zone 6 of Carter and Hart (1977).

Stoliczkaia dispar Zone (lower part): probably *Mortoniceras rostratum* Subzone
Depth 57.51 to 67.00 m

Very silty beds extend down to 67 m, a depth which coincides with an important microfaunal change, characterised by the influx of the planktonic foraminifer *Globigerinelloides bentonensis* in abundance (Professor M. B. Hart, personal communication; Harris, 1982). This phenomenon is seen in Bed XII at Folkestone, which Price (1976; 1977) placed in the topmost *auritus* Subzone following Owen (1976). However, Dr Owen (personal communication) now considers the matrix of Bed XII to be of *rostratum* Subzone age, most of the ammonite fauna present being in the form of black phosphate and derived from the *auritus* Subzone (Owen, 1976, p.491; Gallois and Morter, 1982). The base of the *dispar* Zone is therefore taken at the incoming of this important foraminifer, which is a significant event throughout Northern Europe (Price, 1976; 1977).

The microfaunas are accompanied by macrofossils dominated by the bivalve *Aucellina* down to 59 m. The remainder of the macrofauna includes the bivalves *Chlamys* ex gr. *robinaldina*, *Entolium orbiculare*, *Neithea (Neithella) sp.*, *Plicatula sp.*, '*Pseudolimea*' *sp. juv.* and *Pycnodonte (Phygraea) vesicularis*, together with the cirripede *Arcoscalpellum comptum*. Dr A.W. Medd reports that the nannofossil *Eiffellithus turriseiffeli* is present throughout the *dispar* Zone.

Callihoplites auritus Subzone
Depth 67.00 to 96.70 m

Silty mudstones in the upper part grade down to pale grey mudstones in the lower part. Macrofossils are sparsely represented except below 91 m where a shelly fauna with ammonites is present. An horizon of *Chondrites* is of note at 76 m. No significant changes in the microfaunas are reported by Professor Hart (personal communication) and Harris (1982). Dr Medd reports that *Eiffellithus turriseiffeli* ranges upwards from 71 m; an earlier form of eiffellithid precedes this, nominally called proto-*Eiffellithus*, and ranges upwards from 96 m; the incoming of the latter provides a useful marker for the base of this subzone.

The macrofauna of the *auritus* Subzone includes the coral '*Trochocyathus*' *conulus* (Michelin *non* Phillips); the bivalves *Entolium orbiculare*, *Nucula (Pectinucula) sp.*, *Oxytoma* ex gr. *pectinatum*, *Plicatula*

radiola gurgitis, *Pycnodonte (Phygraea) vesicularis*, together with large chips of *Inoceramus* probably representing '*Inoceramus*' *lissa*, a characteristic fossil for the lower part of the *auritus* and topmost *varicosum* subzones; the ammonites *Callihoplites* cf. *leptus* (at 72.90 m), *Epihoplites sp.*, *Euhoplites* cf. *vulgaris*, *Hysteroceras binum*, *H.* cf. *subbinum*, *Idiohamites spinulosus*, *Mortoniceras sp.*, *Prohysteroceras (Goodhallites) candollianum* Spath *non* Pictet; the crinoid *?Isocrinus legeri*, the echinoid *Hemiaster minimus* (? = *H. maccoyi*), the crab *Notopocorystes (N.) stokesii serotinus* and abundant fish remains.

Hysteroceras varicosum Subzone (upper part)
Depth 96.70 to 106.20 m

These beds are fossiliferous grey mudstones and probably represent a much expanded basal part of Bed XI at Folkestone, which was placed in an expanded *varicosum* Subzone by Owen (1976), thus modifying the original assemblage zone sense of Spath (1923), and making the *auritus* Subzone a partial range zone of *Callihoplites*. The top of the *varicosum* Subzone is now generally recognised by the extinction of epihoplitid ammonites and the entry of the coccolith proto-*Eiffellithus*.

This upper part of the *varicosum* Subzone apparently does not contain the subzonal index (Gallois and Morter, 1982; Owen, 1979; 1984). There is a rich macrofauna, dominated by ammonites, together with corals, bivalves, gastropods and echinoids. The corals include *Cyclocyathus fittoni* at 103 m. The bivalves are dominated by '*Inoceramus*' of the *comancheanus* group with nuculids, oysters, (including *Pycnodonte (Phygraea) vesicularis*) and *Ludbrookia tenuicosta*. Amongst the gastropods *Perissoptera marginata* is common, especially around 101 m; cerithiids (*Nerineopsis sp.*) occur at 102 m; other gastropods include fusinids, trochids and *Ringinella inflata*. The ammonites are dominated by species of *Prohysteroceras*, *Prohysteroceras (Goodhallites)*, *Hysteroceras*, *Euhoplites* and the heteromorph *Idiohamites*.

Some of the species, notably *Prohysteroceras* cf. *candollianum* and *Hysteroceras bucklandi*, were considered by Spath (1934) to be typical of the *auritus* Subzone, but can now be regarded as highest *varicosum* Subzone (Owen, 1976).

Hysteroceras varicosum Subzone (lower part)
Depth 106.20 to 112.48 m

These beds mainly consist of medium grey shelly mudstones which are strongly bioturbated towards the base, and are capped by a phosphatic nodule bed probably representing a non-sequence; a further nodule bed at 111.30 m also indicates a minor non-sequence. The rich shelly fauna consists mainly of ammonites, principally species of *Hysteroceras*, *Epihoplites*, *Semenovites*, *Mortoniceras (Deiradoceras)* and *Anahoplites picteti*, accompanied by species of *Euhoplites* and the heteromorph *Idiohamites*. Fossils typical of Bed X are in the majority, but the rarity of *Hysteroceras varicosum* and the presence of *Anahoplites costosus* and *Hengestites applanatus* is unusual. *Hengestites applanatus* has previously only been recorded from the *auritus* Subzone (see Casey, 1960a), but is now known to occur quite commonly in the upper *varicosum* Subzone. Dr R.Casey (personal communication) suggests that this unit may include strata not represented at Folkestone between beds X and XI.

The corals include *Cyclocyathus fittoni*; the brachiopods include *Kingena spinulosa* and *Biplicatoria sp.* [*Moutonithyris dutempleana* Auctt.]; the bivalves *Birostrina* cf. *concentrica* (common), *Callicymbula phaseolina*, '*Eopecten*' *studeri*, *Atreta sp.*, *Barbatia marullensis*, '*Inoceramus*' cf. *crippsi*, *Nucula (Leionucula) albensis* and *Nucula (Pecti-*

nucula) pectinata. Common gastropods are *Perissoptera marginata* and *Ringinella inflata.* The crinoid *Nielsenicrinus cretaceus* is especially common towards the base of the subzone. Also identified were the echinoid *Hemiaster minimus* and the belemnoid *Neohibolites minimus minimus,* the latter low in the subzone. The base of the *varicosum* Subzone is close to an important microfaunal change recognised by Harris (1982: base of his Zone Diii) and Price (1977: base of his Zone 6), which may indicate a non-sequence and a widespread period of erosion.

Hysteroceras orbignyi Subzone
Depth 112.48 to 117.89 m

These beds consist of medium grey and pale grey very shelly mudstone, with a few muddy limestones and phosphate nodules. The most common shells are the bivalve *Birostrina sulcata* and the related *Birostrina subsulcata,* which are present at the top of the subzone; the latter is taken at the highest occurrence of these two species. The base of the subzone coincides with a slight lithological change, above which the ammonite *Euhoplites inornatus* comes in abundantly.

Corals include *Cyclocyathus fittoni.* Among the bivalves accompanying the dominant *Birostrina sulcata* are *Acila (Truncacila) bivirgata, Placunopsis pseudoradiata, Astarte sp. juv., 'Inoceramus' cf. anglicus, Oxytoma ex gr. pectinatum, Paraesa? regularis, Pycnodonte sp., Ludbrookia tenuicosta, Mesosaccella woodsi.* The gastropod *Cirsocerithium subspinosum* is abundant at 117.13 m and *Semisolarium moniliferum* is common at 117.31 m. The ammonites are mostly species of *Hysteroceras,* notably *H. carinatum, H. orbignyi* and *Hamites spp.;* others include *Dipoloceras* aff. *bouchardianum* trans. to *Mortoniceras, Euhoplites subcrenatus, E. trapezoidalis, E. inornatus* (very common at the base of the zone), *E. ochetonotus, Hamites intermedius distinctus* and *H. intermedius.*

An important microfaunal change has been recognised by Harris (1982: base of his Zone Dii) at about 117 m, which marks the incoming of the benthonic foraminifera *Arenobulimina chapmani.* This is also the base of Price's Zone 5 (Price, 1977). A notable change of fauna occurs within the subzone at a shell bed at 114.85 m, where a generally 'Lower Gault' fauna is replaced by an 'Upper Gault' fauna; here species of *Hamites* give way to species of *Idiohamites* and the echinoid *Hemiaster minimus* is not seen below.

Dipoloceras cristatum Subzone
(a) Depth 117.89 to 125.20 m

These beds comprise medium to dark grey and medium to pale grey mudstones with a shelly fauna and *Chondrites.*

The base of the beds is taken at a phosphatic nodule bed at 125.20 m, which probably marks a non-sequence. This break defines the base of the Di microfossil Zone of Harris (1982), but the major microfaunal event occurs above this at 124 m where *Citharinella pinnaeformis* enters, marking the base of Zone 4a of Carter and Hart (1977) and Zone 4i of Price (1977).

The macrofaunas are distinctive and include the coral *Cyclocyathus fittoni* and bivalves dominated by *Birostrina sulcata,* with *Birostrina subsulcata* and *Birostrina concentrica gryphaeoides* as well as *Acila (Truncacila) bivirgata, Ensigervilleia forbesiana* common around 119 m, *Corbula gaultina* common around 121 m, *Cymbula vibrayeana, Longinuculana solea* common, *Ludbrookia tenuicosta, Mesosaccella woodsi* common throughout, *Nucula (Pectinucula) pectinata* common, *Nucula (Leionucula) ovata* and *Nucula gaultina.* Apart from *Perissoptera spp.,* the most notable gastropod occurrence in this subzone is the abundance of *Cirsocerithium spinulosum* around 120 m. The opisthobranch *Avellana dupiniana* is also present. The scaphopods *Dentalium (Fissidentalium) decussatum* (common) and *Dentalium (Laevidentalium) jefferysi* were identified; the latter appears to be restricted to this subzone. The ammonites were *Anahoplites planus, Beudanticeras* cf. *subparandieri, Euhoplites* aff. *inornatus, Euhoplites ochetonotus, Metaclavites compressus* and *Hysteroceras sp.* at 122.04 m. Other fossils include fish

remains and a tooth of *Lamna appendiculata.*
(b) Depth 125.20 to 126.55 m (age uncertain)
These beds comprise pale grey extensively bioturbated mudstone with *Chondrites.* The fauna is very limited and consists mainly of the bivalve *Birostrina subsulcata* in the upper part and small *Birostrina concentrica.* Only long-ranging ammonites are present and the age of these beds is in some doubt; *Anahoplites planus discoideus, A. sp.,* and an undetermined brancoceratid ammonite occur.

These beds probably belong to the lower part of the *cristatum* Subzone as seen at Wissant. Owen (1971, p.126) took the base of the latter at the lower of two nodule beds in the Aycliff Borehole, the representatives of which occur in the Glyndebourne Borehole at 124.66 and 125.20 m, but both boreholes show that *B. subsulcata* is represented below the lower nodule bed and this is an indication that the lower *cristatum* Subzone is present (Owen, 1976).

Harris (1982) has suggested that this unit may be of *daviesi* Subzone age on the basis of ostracod identification, but his evidence is not conclusive and the *daviesi* Subzone may well be absent. These beds would seem to fall within the foraminiferal Zone 4 of Carter and Hart (1977) and Zone 4 of Price (1977) and Zone Cii of Harris (1982), though they lie above the topmost occurrence of the ostracod *Cytherelloidea chapmani.*

?Euhoplites nitidus Subzone
Depth 126.55 to 128.34 m

These beds consist of medium to dark grey mudstone with *Chondrites.* A limited and non-diagnostic assemblage of fossils was collected, chiefly *Birostrina concentrica.* The base of the subzone is drawn at 128.34 m, above the occurrence of *Euhoplites loricatus,* which relates to the *meandrinus* Subzone (Owen, 1971). The ammonites are mostly *Anahoplites* or *Dimorphoplites,* including *D. tethydis,* and the heteromorph *Hamites tenuis.* The bivalves include *Entolium orbiculare, Longinuculana solea, Nucula (Leionucula) ovata* and *Nucula (Pectinucula) pectinata,* and the gastropods *Perissoptera carinata* and *Cirsocerithium subspinosum.* The scaphopod *Dentalium (Fissidentalium) decussatum,* the coral *Cyclocyathus fittoni,* several examples of *Hemiaster bailyi* and crustacean remains are also present. The base of the subzone almost coincides with an important microfaunal change, the base of Zone Cii of Harris (1982), and the base of Zone 4 of Price (1977).

Euhoplites meandrinus Subzone
Depth 128.34 to 131.13 m

These beds are lithologically similar to those above, but the mudstones are paler grey. The ammonites include *Euhoplites loricatus,* fragments of *Dimorphoplites* and *Hamites spp.,* including *H. attenuatus,* the last occurring towards the base of the subzone. The bivalves are *Entolium orbiculare* (common), *Nanonavis carinata, Nucula (Pectinucula) pectinata, Thracia sanctae-crucis.* Also noted were the gastropods cf. *Perissoptera spp.*

Mojsisovicsia subdelaruei Subzone
Depth 131.13 to 131.33 m

These beds consist of medium to pale grey mudstone with *Chondrites.* This subzone is recognised by the presence of *Mojsisovicsia sp.* and is typically thin in Sussex (see Owen, 1971). *Birostrina concentrica* also occurs. The subzone marks the base of foraminiferal Zone 4 of Carter and Hart (1977), with the incoming of *Hoeglundina carpenteri,* together with an abundance of robertinacean foraminifera such as *Epistomina spinulifera.*

Dimorphoplites niobe Subzone
Depth 131.33 to 132.54 m

These beds comprise medium to pale grey mudstone with buff clay-ironstone in the lower part. This subzone appears to be unusually thin, although the fauna is not diagnostic. Ammonites include *Dimorphoplites sp.* and *Hamites attenuatus.* Bivalves are dominated by *Birostrina concentrica* and many *Nucula (Pectinucula) pectinata.* The

'Crab Bed' of Bed III at Folkestone is represented by examples of *Notopocorystes*. The base of the subzone coincides with an important microfaunal event and represents the base of Zone Ci of Harris (1982) and of Zone 3iv of Price (1977).

Anahoplites intermedius Subzone
Depth 132.54 to 147.24 m

These beds consist of medium to pale grey mudstone with a variable mica content, which become brownish in colour towards the base. This would appear to be a thick development of the subzone. The ammonites are dominated by species of *Anahoplites* and *Dimorphoplites*; horizons of abundant *Falciferella milbournei*, which is typically confined to this subzone (Casey, 1954) also occur. They are cf. *Anahoplites grimsdalei*, *A. intermedius* (common), *A. mantelli* (common), *A. planus*, *Anahoplites praecox*, cf. *Dimorphoplites alternans perelegans*, *Falciferella milbournei*, *Hamites attenuatus* and *H. sp.* The bivalves are dominated by *Birostrina concentrica* throughout, and by *'Anomia' carregozica* down to 135.60 m. The latter is commonly found in Bed II at Folkestone (see range chart of Price, 1879). Other bivalves include *Placunopsis* cf. *pseudoradiata*, *Corbula gaultina*, *Cymbula ?vibrayeana*, *Entolium orbiculare*, *Ensigervilleia forbesiana*, *Longinuculana solea*, *Mesosaccella woodsi*, *Modiolus sp.*, *Nucula (Leionucula) sp.*, *Nucula (Pectinucula) pectinata*, *Pseudolimea gaultina* and *Pseudoptera sp.* Among the gastropods, aporrhaids are most common and include the characteristic Bed II species *Drepanocheilus doratochila* throughout most of the subzone, together with common *Perissoptera carinata* and *P. marginata*; also noted were *Actaeon affinis* (common), *Drepanocheilus carinella*, *Avellana* cf. *dupiniana*, *A. pulchella*, cf. *Cirsocerithium subspinosum* and *Rissoina sowerbyii* (common). Other fossil remains include the annelid *Parsimonia sp.*, the coral *Cyclocyathus fittoni* and a small undescribed caryophylliid. The echinoid *Hemiaster sp.*, together with many fish remains including a shark's tooth and fish otolith, are also present. An important microfaunal event occurs at 137 m, where the ostracod *Dolocytheridea bosquetiana* appears. This would appear to be a consistent marker for the 'middle' of the *intermedius* Subzone, bed 4 in East Anglia (Wilkinson and Morter, 1981; Gallois and Morter, 1982). The first planktonic foraminifera were identified at 142 m, though they have been recorded in the early Albian at Folkestone (Hart, 1973).

Hoplites spathi Subzone (Owen, 1971)
Depth 147.24 to 148.65 m

This subzone is equivalent to the *Hoplites dentatus–spathi* Subzone (Casey *in* Smart and others, 1966, pp.102–113). The beds comprise medium grey to brown smooth mudstone with little mica; darker grey mudstone occurs towards the base. The boundary with the subzone above is taken at a pale nodule bed, below which *Hoplites (Hoplites) spp.* are abundant.

The ammonites are *Hoplites (H.)* cf. *dentatus*, *Hoplites (H.)* cf. *spathi*, *Hoplites (H.) spp.* The belemnite *Neohibolites minimus minimus* is common and the bivalves are dominated by *Birostrina concentrica*, with *Ludbrookia tenuicosta*, *Mesosaccella woodsi* and *Placunopsis* cf. *pseudoradiata*. Other fossils include small caryophylliid corals and abundant fish remains.

Lyelliceras lyelli Subzone: *Hoplites benettianus* Subzone
Depth 148.65 to at least 149.00 m

These beds comprise dark grey mudstone which contains much glauconite, and they grade downwards into increasingly sandy clays. The boundary with the *spathi* Subzone above is sharp, probably representing a non-sequence. The fauna is very limited, with some indeterminate mould fossils and fish remains, but includes well-preserved phosphatic ammonites identified as *Beudanticeras laevigatum*, *Lyelliceras pseudolyelli* and hamitid indet. Below 149 m palaeontological evidence is lacking. The base of the Gault was proved at 152.60 m.   AAM

APPENDIX 4

List of Geological Survey photographs

The more recent photographs are listed below. Copies of these are deposited for reference in the British Geological Survey library at Keyworth, Nottingham NG12 5GG. Prints and slides can be supplied at a fixed tariff. The photographs were taken by Messrs C. J. Jeffery and J. M. Pulsford. All numbers belong to Series A.

9936   Cement works utilising chalk, Southerham.
9941   The Brooks and the South Downs, near Rodmell cement works, Lewes.
9942   Firle Beacon and the South Downs, from near Glynde.
9947   Glynde Gap viewed from near Kingston.
9952   Downstream view of the Cuckmere valley near Alfriston.
12272   Weald Clay, Chailey brick-pit.
12273   Flinty Head, Chailey brick-pit.
12274   Barn constructed of Large-'Paludina' limestone, near Laughton.
12275   Detail of limestone walling material (in above).
12276   View of Weald Clay vale between Ripe and Laughton.
12277   Tunbridge Wells Sand in pit at Burgh Hill.
12278   Dip-slope of the Tunbridge Wells Sand near Muddles Green.
12279   Tunbridge Wells Sand in pit near Hellingly Hospital.
12280   Tunbridge Wells Sand scenery near Hellingly.
12281   View of Pevensey Levels from Church Farm, Herstmonceux.
12282   Red clay horizon in Weald Clay, Keymer brick-pit.
12283–4   Chalk scarp scenery, near Folkington.
12285   Section in Lower and Middle Chalk, Ash Chalk Pits, Jevington.
12286   Pevensey Levels, from Rickney.
12287   Silty clays in Tunbridge Wells Sand in a pit at Glynleigh.
12288   Tunbridge Wells Sand in a pit at Glynleigh.
12289–90   Pevensey Castle.

12291   Wall stones of Pevensey Castle.
12292   Pevensey Levels and Herstmonceux ridge from near Newhouse.
12293–4   Top Ashdown Sandstone in sunken lane near Waldron.
12295   Bell-pits for ironstone in Wadhurst Clay near Horam.
12296   Hastings Beds scenery near Horam.
12297   Top Ashdown Sandstone and Wadhurst Clay feature near Horam.
12298–9   Wadhurst Clay, Marle Green brick-pit.
12300   Valley-bulge in Wadhurst Clay near Vine's Cross.
12301   Ashdown Beds in bank of stream near Warbleton.
12302   Large-'Paludina' limestone feature near Laughton.
12303–4   Laughton Levels near Laughton.
12305   Exposure of Folkestone Beds sands, Selmeston.
12306   Chalk escarpment from Sherman Bridge.
12307   Upper Chalk scenery, Beddingham Hill.
12308   Dry valleys south of Firle Beacon.
12309   Gault and Lower Chalk scenery from Firle Beacon.
12310   Lower Chalk, Newington's pit, Glynde.
12311   Lithologies of Lower Chalk, Newington's pit, Glynde.
12312   Clay-with-flints in an exposure near Bopeep.
12313   South Downs scarp and The Brooks from Beddingham.
12314   Gault in pit at the former Rodmell cement works.
12315   Section in Lower Chalk at the former Rodmell cement works.
12316   Coombe Deposits in a section at the former Rodmell cement works.
12317   Dressed flints as building stone, Old Grammar School, Lewes.
12318   Upper Chalk in the Caburn Syncline, Southerham.
12319   The Brooks, Lewes, from near Iford.
12320   The Brooks, Lewes, from near Rodmell.
12321   Ouse valley near Lewes.
12322   Ardingly Sandstone in a pit near 'The Rocks', Uckfield.
12323   Ardingly Sandstone in a sunken lane near Newick.
12324   Ouse valley and chalk escarpment north of Lewes.

INDEX OF FOSSILS

GENERAL INDEX

BRITISH GEOLOGICAL SURVEY

Keyworth, Nottingham NG12 5GG
Plumtree (060 77) 6111

Murchison House, West Mains Road,
Edinburgh EH9 3LA   (031) 667 1000

The full range of Survey publications is available through the Sales Desks at Keyworth and Murchison House. Selected items are stocked by the Geological Museum Bookshop, Exhibition Road, London SW7 2DE; all other items may be obtained through the BGS London Information Office in the Geological Museum ((01) 589 4090). All the books are listed in HMSO's Sectional List 45. Maps are listed in the BGS Map Catalogue and Ordnance Survey's Trade Catalogue. They can be bought from Ordnance Survey Agents as well as from BGS.

The British Geological Survey carries out the geological survey of Great Britain and Northern Ireland (the latter as an agency service for the government of Northern Ireland), and of the surrounding continental shelf, as well as its basic research projects. It also undertakes programmes of British technical aid in geology in developing countries as arranged by the Overseas Development Administration.

The British Geological Survey is a component body of the Natural Environment Research Council.

Maps and diagrams in this book use topography based on Ordnance Survey mapping

HER MAJESTY'S STATIONERY OFFICE

HMSO publications are available from:

HMSO Publications Centre
(Mail and telephone orders)
PO Box 276, London SW8 5DT
Telephone orders (01) 622 3316
General enquiries (01) 211 5656
Queueing system in operation for both numbers

HMSO Bookshops
49 High Holborn, London WC1V 6HB
    (01) 211 5656 (Counter service only)
258 Broad Street, Birmingham B1 2HE
    (021) 643 3740
Southey House, 33 Wine Street, Bristol BS1 2BQ
    (0272) 264306
9 Princess Street, Manchester M60 8AS
    (061) 834 7201
80 Chichester Street, Belfast BT1 4JY
    (0232) 238451
71 Lothian Road, Edinburgh EH3 9AZ
    (031) 228 4181

HMSO's Accredited Agents
(see Yellow Pages)

And through good booksellers